THE PYRENOMYCETOUS FUNGI

LEWIS E. WEHMEYER †

MYCOLOGIA MEMOIR NO. 6

Edited by: Richard T. Hanlin,
Department of Plant Pathology,
University of Georgia, Athens, Georgia

Published for
The New York Botanical Garden
In Collaboration With
The Mycological Society of America

J. CRAMER PUBLISHER
P.O. Box 48,
3301 LEHRE GERMANY
1975

LIBRARY OF CONGRESS CATALOG CARD NUMBER 75-29771

Printed in Germany
28.XI.1975
by Strauss & Cramer GmbH, D-6901 Leutershausen
ISBN 3-7682-0967-9

LEWIS E. WEHMEYER

LEWIS EDGAR WEHMEYER*

1897 - 1971

Lewis E. Wehmeyer was born January 1, 1897 in Quincy, Illinois. He was the son of August H. and Emma (Grimm) Wehmeyer. His early years were spent in Quincy where he attended elementary and secondary schools, graduating from Quincy High School in 1914. During these early years he acquired an interest in biology and in 1916 he entered the University of Michigan to further his education. His studies were interrupted by World War I, and he served as a sergeant in the U.S. Army Corps of Engineers during 1918. He returned to Ann Arbor and received the Bachelor of Science degree in Forestry in 1921. He then undertook graduate studies in mycology at Michigan under Prof. Calvin H. Kauffman, receiving the Doctor of Philosophy degree in Botany in 1925. His last year at Michigan was spent as an Emma Cole Fellow. Following graduation Dr. Wehmeyer spent three years at Harvard University on a National Research Fellowship, where he taught and worked with Prof. Wm. H. Weston. During his stay at Harvard he became a close friend of Prof. A.R. Prince, whom he accompanied on a collecting trip to Nova Scotia in 1926. It was during a visit at the Prince home that he met Elaine Prince; they were married on September 1, 1927.

In 1928 Dr. Wehmeyer returned to the University of Michigan as an instructor in botany. He became assistant professor in 1931, associate professor in 1937, and full professor in 1947. Upon his retirement in 1967 he was named Emeritus Professor. After retirement he continued to work as much as his health would permit, writing a comprehensive treatment of the perithecial Ascomycetes. Dr. Wehmeyer was a member of various scientific societies, including the American Association for the Advancement of Science, Sigma Xi, Mycological Society of America, Gamma Alpha, International Association for Plant Taxonomy, American Society of Plant Taxonomists, Phi Sigma, Botanical Society of America, Michigan

---

*By Richard T. Hanlin, University of Georgia, Athens.

Academy of Science, Arts and Letters, and the University of Michigan Research Club. He died September 11, 1971. He is survived by Elaine, his devoted wife of 44 years. There were no children.

During his more than forty years as a mycologist, Dr. Wehmeyer made significant contributions to our knowledge of the pyrenomycetous Ascomycetes. His early studies were devoted to the stromatic pyrenomycetes and made him an authority on these fungi. It was during these studies that he coined the term "centrum" for the internal tissues of perithecial ascocarps, replacing the German "nukleus" which was used at that time. He then published a long series of papers on the life histories of diaporthaceous fungi which greatly increased our knowledge of this group. These studies culminated in monographs of *Diaporthe* and related genera. During this period too he published a series of floristic papers on Nova Scotia and neighboring areas, and of Wyoming. It was during these studies that he became interested in the genus *Pleospora*, which was to occupy him for the next 15 years, culminating in a world monograph of the genus. While working on *Pleospora* and related genera he published several papers on the developmental morphology of these fungi. In recognition of his work, *Pleospora wehmeyeri* and *Didymella wehmeyeri* were named in his honor.

Dr. Wehmeyer was an avid collector and traveller. His first trips to the Western United States were made with Dr. C.H. Kauffman while he was still a graduate student. During the ensuing years he made numerous trips to the western states and Canada. Through the years he also visited many herbaria around the world to study specimens. Among the countries he worked in were England, Sweden, Denmark, Italy,Germany, Czechoslovakia, Austria, France, Portugal, Spain, Switzerland, Belgium, the Netherlands, and Singapore. As a result of these trips he personally knew many of the world's leading mycologists.

In addition to his research Dr. Wehmeyer taught elementary botany, introductory mycology, and a course in Ascomycetes and Fungi Imperfecti. He also served as major professor to a number of students, several of whom have gone on to make significant contributions of their own.

When not collecting fungi or travelling Dr. Wehmeyer was usually involved with his hobbies of gardening and stamp collecting. He also enjoyed spending summers at his cottage in Seal Cove, Maine. During his later years his work was curtailed somewhat by several serious illnesses, the last of which resulted in his death.

I first became acquainted with Dr. Wehmeyer just over 20 years ago when I enrolled in his introductory mycology course. As a student I was impressed not only with his broad knowledge of the fungi, but also with his great enthusiasm for them. This was perhaps best shown in the field, as he eagerly collected everything in sight, exclaiming over some rare find. It was this enthusiastic introduction to mycology that later led me to pursue a similar career. It was also the beginning of a warm personal friendship that continued until his death.

BIBLIOGRAPHY OF LEWIS EDGAR WEHMEYER

1924

Wehmeyer, L. E. 1924. The imperfect stage of some higher Pyrenomycetes obtained in culture. Pap. Mich. Acad. **3**: 245-266, *pl. 21-23.*

1925

——. 1925. The perfect stage of the Valsaceae in culture and the hypothesis of sexual strains in this group. Pap. Mich. Acad. **4**: 395-412, *pl. 19.*

1926

——. 1926. Cultural life-histories of certain species of *Eutypella, Diatrypella* and *Cryptovalsa.* Pap. Mich. Acad. **5**: 179-194, *pl. 18.*

——. 1926. Further cultural life histories of the stromatic Sphaeriales. Amer. J. Botany **13**: 231-247, *pl. 12-15.*

——. 1926. Cultural life histories of *Melanconis* and *Pseudovalsa.* Mycologia **18**: 257-273, *pl. 30-33.*

——. 1926. A biologic and phylogenetic study of the stromatic Sphaeriales. Amer. J. Botany **13**: 575-645.

1927

——. 1927. Cultural life-histories of *Diaporthe* I. Pap. Mich. Acad. **6**: 377-396, *pl. 23-26.*

——. 1927. Cultural life histories of *Diaporthe.* II. Mycologia **19**: 165-182, *pl. 13-15.*

1928

——. 1928. Cultural life-histories of *Diaporthe* III. Pap. Mich. Acad. **8**: 215-232, *pl. 15-17.*

1929

——. 1929. Cultural life histories of *Diaporthe* IV. Pap. Mich. Acad. **9**: 477-494, *pl. 87-89.*

1930

——. 1930. Studies on the role of the host in the genus *Diaporthe.* Pap. Mich. Acad. **11**: 309-328, *pl. 34-36.*

1933

——. 1933. The British species of the genus *Diaporthe* Nits. and its segregates. Trans. Brit. Mycol. Soc. **17**: 237-295.

——. 1933. The genus *Diaporthe* Nitschke and its segregates. Univ. Mich. Studies, Sci. Ser. 9. × + 349 p., *18 pl.*

1935

Jenkins, Anna E., and L. E. Wehmeyer. 1935. Transfer of *Diaporthe umbrina* to the genus *Cryptosporella*. Phytopathology **25**: 886-889, *1 fig.*

Wehmeyer, L. E. 1935. Contributions to a study of the fungous flora of Nova Scotia. I. Pap. Mich. Acad. **20**: 233-266, *pl. 51-52.*

1936

——. 1936. Cultural studies of three new Pyrenomycetes. Mycologia **28**: 35-46, *3 fig.*

Smith, A. H., and L. E. Wehmeyer. 1936. Contributions to a study of the fungous flora of Nova Scotia II. Agaricaceae and Boletaceae. Pap. Mich. Acad. **21**: 163-197, *pl. 26-27.*

Wehmeyer, L. E., and K. Cejp. 1936. Contributions to a study of the fungous flora of Nova Scotia III. Resupinate Hydnaceae. Pap. Mich. Acad. **21**: 209-213.

Wehmeyer, L. E. 1936. Cultural life histories of *Melanconis* and *Pseudovalsa.* II. Mycologia **28**: 528-541, *5 fig.*

1937

——. 1937. Studies of certain species of *Melanconis* on *Carpinus, Ostrya* and *Corylus.* Mycologia **29**: 599-617, *3 fig.*

1938

——. 1938. Cultural life histories of *Melanconis* and *Pseudovalsa*. III. The genus *Prosthecium*. Fres. emend. Petrak. Pap. Mich. Acad. **23**: 233-242, *3 pl.*

——. 1938. Las especies de "*Diaporthe*" en el Herbario Spegazzini. Rev. Museo La Plata, N.S., **2**: 65-88, *3 pl.*

——. 1938. Addenda a las "*Diaporthe*" del Herbario Spegazzini. Notas Museo La Plata **3**: 99-102.

1939

——. 1939. The genus *Massariovalsa*. Amer. J. Botany **26**: 831-834, *9 fig.*

1940

——. 1940. Contributions to a study of the fungus flora of Nova Scotia. IV. Additional Basidiomycetes. Canad. J. Res., C, **18**: 92-110, *1 fig., 1 pl.*

——. 1940. Contributions to a study of the fungus flora of Nova Scotia. V. Discomycetes. Canad. J. Res., C, **18**: 535-549, *8 fig.*

——. 1940. Cultural life histories of *Melanconis* and *Pseudovalsa.* IV. Mycologia **32**: 321-330, *22 fig.*

1941

——. 1941. A revision of *Melanconis, Pseudovalsa, Prosthecium,* and *Titania.* Univ. Mich. Studies, Sci. Ser. 14. viii + 161 p., *11 pl.*

——. 1941. *Pseudotrichia* and the new genus *Phragmodiaporthe.* Mycologia **33**: 54-63, *11 fig.*

——. 1941. The genus *Thyridaria* (Pyrenomycetes). Lloydia **4**: 241-261, *4 pl.*

1942

——. 1942. Contributions to a study of the fungous flora of Nova Scotia. VI. Pyrenomycetes. Canad. J. Res., C, **20**: 572-594, *18 fig.*

1946

——. 1946. Studies on some fungi from northwestern Wyoming. I. Pyrenomycetes. Mycologia **38**: 144-170, *20 fig.*

——. 1946. Studies on some fungi from northwestern Wyoming. II. Fungi Imperfecti. Mycologia **38**: 306-330, *28 fig.*

——. 1946. Studies on some fungi of northwestern Wyoming. III. *Pleospora* and *Leptosphaeria.* Lloydia **9**: 203-240, *3 pl.*

1947

——. 1947. Studies on some fungi from northwestern Wyoming. IV. Miscellaneous. Mycologia **39**: 463-478, *2 fig.*

——. 1948. The developmental pattern within the genus *Pleospora* Rab. Mycologia **40**: 269-294, *56 fig.*

1949

——. 1949. (Review of) Biology of Pathogenic Fungi, ed. by Walter J. Nickerson. Mycologia **41**: 95-96.

——. 1949. Studies in the genus *Pleospora*. I. Mycologia **41**: 565-593, *17 fig.*

1950

——. 1950. The fungi of New Brunswick, Nova Scotia, and Prince Edward Island. Ottawa, National Research Council of Canada. vii + 150 p.

——. 1950. Considerations of some methods of approach to the taxonomic study of the fungi. Proc. Seventh Intern. Bot. Congr. (Stockholm), p. 393-394.

1951

——. 1951. Studies in the genus *Pleospora*. III. Mycologia **43**: 34-53, *23 fig.*

——. 1951. Studies in the genus *Pleospora*. IV. Mycologia **43**: 570-589, *20 fig.*

——. 1951. (Review of) Researches on Fungi: The Sexual Process in the Uredinales, Vol. VIII, by A. H. Reginald Buller. Science **113**: 455-456.

1952

——. 1952. Studies in the genus *Pleospora*. VI. Amer. J. Botany **39**: 237-243, *13 fig.*

——. 1952. Studies in the genus *Pleospora.* II. Lloydia **15**: 65-109, *61 fig.*

——. 1952. The genera *Leptosphaeria, Pleospora,* and *Clathrospora* in Mt. Rainier National Park. Mycologia **44**: 621-655, *24 fig.*

——. 1952. The collection and preservation of ascomycetous fungi. Asa Gray Bull., N.S., **1**: 315-323.

——. 1952. Some pryenomycetous fungi from Mt. Rainier National Park. Sydowia **6**: 412-428, *1 fig., pl. 7.*

——. 1952. Some Fungi Imperfecti from Mt. Rainier National Park. Sydowia **6**: 433-442, *pl. 8.*

1953

——. 1953. Studies in the genus *Pleospora*. V. Mycologia **45**: 391-414, *18 fig.*

——. 1953. On the status of the generic names *Pyrenophora* and *Pleospora*. Mycologia **45**: 562-571.

1954

——. 1954. Studies in the genus *Clathrospora*. Mycologia **46**: 498-522, *19 fig.*

——. 1954. Perithecial development in *Pleospora trichostoma*. Bot. Gaz. **115**: 297-310, *52 fig.*

Müller, E., and L. E. Wehmeyer. 1954. Ein neuer, alpiner Ascomycet: *Mycosphaerella phacae-frigidae*. Sydowia **8**: 190-191, *1 fig.*

1955

Wehmeyer, L. E. 1955. The development of the ascocarp in *Pseudoplea gaeumannii*. Mycologia **47**: 163-176, *33 fig.*

——. 1955. Development of the ascostroma in *Pleospora armeriae* of the *Pleospora herbarum* complex. Mycologia **47**: 821-834, *27 fig.*

1957

——. 1957. The genera *Saccothecium, Pringsheimia, Pleosphaerulina* and *Pseudoplea*. Mycologia **49**: 83-94, *7 fig.*

——. 1957. (Review of) Danish Pyrenomycetes, by Anders Munk. Mycologia **49**: 906.

——. 1957. (Review of) Études Taxonomiques sur les Pléosporacées, by Lennart Holm. Mycologia **49**: 906-907.

——. 1957. The genus *Montagnula* Berl. Sydowia Beih. **1**: 257-263, *pl. 11-12.*

1959

——. 1959 *Leptosphaerulina* McAlp. antedates *Pseudoplea* Höhn. Sydowia **12**: 490-491.

1960

——. 1960. (Review of) Biogeography: An Ecological Perspective, by Pierre Dansereau. Mycopath. Mycol. Appl. **12**: 169-170.

1961

——. 1961. A world monograph of the genus *Pleospora* and its segregates. Ann. Arbor, University of Michigan Press. ix + 451 p., *25 pl.*

1963

——. 1963. Some Himalayan Ascomycetes of the Punjab and Kashmir. Mycologia **55**: 309-336, *17 fig.*

1964

Wehmeyer, L. E., and S. Ahmad. 1964. Some Ascomycetes of West Pakistan. Biologia **10**: 7-21, *20 fig.*

Wehmeyer, L.E. 1964. Some Fungi Imperfecti of the Punjab and Kashmir. Mycologia **56**: 29-52, *36 fig.*

1966

——. 1966. Development of the ascocarp in *Cryptomycina pteridis.* Mycologia **58**: 752-760, *27 fig.*

EDITOR'S NOTE

Perhaps the most difficult task that can befall an editor is the preparation of a manuscript for posthumous publication. Corrections must be made with care lest they change the original intent of the author. The present manuscript was submitted to the Mycologia Memoirs Committee by Prof. Wehmeyer a few weeks before his death. Although it was in essentially final form it had not received the final editing necessary for publication. Also, line drawings which Prof. Wehmeyer intended to include were not sufficiently complete to be used. Thus the Committee was faced with the difficult decision of how to proceed with the manuscript. At length the decision was made to publish the manuscript as a memorial to Prof. Wehmeyer. Two factors weighed heavily in this decision. First was the plea of Mrs. Wehmeyer that her husbands's final and largest work not go unrecognized, and secondly, the feeling of the Committee that Prof. Wehmeyer's opinions on the relationships of the Ascomycetes would be of value to students of these fungi, since they had not been published previously. This work also represents the only contribution in English in recent years to cover a wide range of Ascomycetes. Some will undoubtedly disagree with this decision, but it is hoped that they will utilize the useful information it does contain.

Since the manuscript could not be published as received there remained the difficult task of preparing it for publication. At the request of Mrs. Wehmeyer I agreed to attempt this, out of respect for my former mentor. In this I owe a debt of gratitude to Dr. Margaret Barr Bigelow, also a former student of Prof. Wehmeyer, for carefully reviewing the manuscript and offering many valuable suggestions. The opinions and comments of Drs. E.S. Luttrell and E.G. Simmons are also greatly appreciated. Changes were kept to a minimum to avoid altering the ideas expressed by Prof. Wehmeyer. The index was prepared by Dr. G.K. Batra and the final manuscript was typed by Dr. R.O. Blanchard. Finally, we are grateful to the National Science Foundation for permission to use funds from Prof. Wehmeyer's Grant GB-5132, "Morphologic and Cytologic Developmental Studies in the Ascomycetes" to prepare the final copy of the manuscript. This was done under the supervision of Dr. R.L. Shaffer, University of Michigan, who also checked the list of publications for accuracy.

Richard T. Hanlin, May, 1973

INTRODUCTION

The Ascomycetes

The Ascomycetes are a large and complex group, They are found everywhere from the arctic to the tropics. They may be parasitic or saprophytic, and are found as parasites on both plants and animals, or as saprophytes in the soil, in fresh and salt water, on all sorts of vegetable debris, on dung and even on the hair and skin of animals and humans and feathers of birds. The literature of the group is vast and scattered. Although a great deal of work has been done on the morphology, culture, physiology, cytology and genetics of the group, they remain the least well known of the great divisions of the fungi. Their taxonomy is in a state of flux and their treatment in text books is for the most part scanty. It would be folly for any one individual to attempt a comprehensive account of the Ascomycetes as a whole. The present account is merely a summation of the major points of our knowledge of the group.

The Ascomycetes are a clearly circumscribed group, with one character common to all of them, namely the formation of an ascus. An ascus may arise in many ways. It is initially a binucleate cell. The two nuclei fuse and then undergo meiosis and usually one or more mitoses. The ascospores are formed by free cell formation: a mass of protoplasm is delimited around each nucleus (or group of nuclei). A typical ascus has eight ascospores, but according to the species, there may be one, two, four or hundreds of ascospores at maturity.

Origin

There have been several theories concerning the origin of the Ascomycetes. De Bary (**1881, 1884, 1887**) suggested their origin from such Phycomycetes as the Mucorales and Peronosporales because of the similar fusion of two gametic structures to form an oospore, zygospore or ascus. Brefeld (**1889-1891**) also derived the Ascomycetes from the Phycomycetes, but supposed the ascus to be a modified sporangium with a fixed number of spores. Present cytological knowledge now discredits these theories. Sachs (**1874, 1896**) included the Florideae (red algae) and the Ascomycetes in the same class, the Carposporeae. Although he based this arrangement merely on the morphologic similarity of the fruiting structures of the two groups, this similarity of the gametic hyphae of certain lichens and some Discomycetes to those of the red algae, was taken up as a theory of the origin of the Ascomycetes from the red algae. This theory was held for a number of years by many investigators (Gwynne-Vaughan, B.O. Dodge, Fraser and others). Much of the earlier work on ascocarp development and cytology was influenced by it. Atkinson (**1912, 1915**) was perhaps the first to deny this theory. He believed the Ascomycetes arose from such Phycomycetes as *Piptocephalis* and the Peronosporales, because of the similarity between the gametangia forming oospores and zygospores and the asci of such genera as *Monascus* and *Gymnoascus*, which he considered to be the closest relatives in the Ascomycetes. In all of these theories the supposed primitive Ascomycetes (Lichens, Operculate Discomycetes) were already well developed with a true mycelium and an advanced type of ascocarp. The Protoascales and yeasts were looked upon as degenerate forms with a simple morphology due to their saprophytic or parasitic existence. In my opinion the logical origin of the Ascomycetes are those very simple taxa with simple unicells or short chains of pseudomycelia as in the Saccharomycetaceae. From this group, which shows a wide variety of evolutionary lines of development, the more complex Ascomycetes can be derived. Such an origin would present a progressive rather than a regressive line of evolution.

Historical

There have been many arrangements of the Ascomycetes proposed throughout the years. Nearly every investigator has presented his modified version, reflecting his individual viewpoint. The earliest arrangements of Persoon (**1801**) and Fries (**1821-1823**) were largely descriptive and entirely artificial, based on the superficial appearance of the ascocarp. Later Fries (**1849**) and Fuckel (**1870**) recognized a number of larger taxa which included related species. It was the systems of Winter (**1887**) and Rehm (**1896**) in Rabenhorst's *Kryptogamenflora* and Lindau (**1897**) in Engler and Prantl's *Die Natürlichen Pflanzenfamilien* which provided a classification with keys to the genera. These systems were also largely artificial and were based upon the gross morphology of the ascocarp. Nevertheless, they served for many years as the basis for determination of the Ascomycetes and most systems presented throughout the early twentieth century were based upon these outlines.

The first radical change in approach started with von Höhnel's (**1907**) erection of the Pseudosphaeriaceae. Fries as early as 1849 had recognized that some pyrenomycetous fungi produced asci in locules within a stroma, rather than in a true perithecial wall. Nitschke (in Fuckel, **1870**) proposed the family Dothideaceae for such stromatic forms. In 1907,

Höhnel reported that certain fungi (related to *Leptosphaeria*) formed only one or a few asci within a spherical stroma. These asci, he said, arose in separate chambers within a common stroma and were separated from one another by stromatic ground tissue which was compressed by the enlargement of the asci into strips of elongated cells attached at both the top and the bottom of the locule. Therefore, he said, they were not true perithecia but small spherical stromata and should be placed next to the Dothideales as a new family, the Pseudosphaeriaceae. There was a great deal of discussion in the years following by Höhnel, Petrak, Theissen, Sydow and others as to how to distinguish a true perithecium from a stromatic ascocarp.

Saccardo, in *Sylloge Fungorum*, **(1882-1931)** divided the major taxa of the Ascomycetes solely on the basis of spore color and septation, as follows:

Spores 1-celled (Amerosporeae); spores hyaline (Hyalosporeae); spores colored (Phaeosporeae)
Spores 2-celled (Didymosporeae); spores hyaline (Hyalodidymeae); spores colored (Phaeodidymeae)
Spores 3-many celled (Phragmosporeae); spores hyaline (Hyalophragmeae); spores colored (Phaeophragmeae)
Spores muriform (Dictyosporeae); spores hyaline (Hyalodicteae); spores colored (Phaeodicteae)
Spores elongate to filiform (Scolecosporeae)
Spores lobed or branched (Staurosporeae)
Spores coiled or a spiral (Helicosporeae)

Nannfeldt (**1932**), in his study of the inoperculate Discomycetes, divided all of the ascocarp forming Ascomycetes into two groups: the Ascoloculares, including those with a stromatic ascocarp, and the Ascohymeniales, those with a true perithecium or apothecium and a hymenial layer of asci. This use of the development of the ascocarp to indicate relationships was taken up by Miller (**1949**) and Luttrell and by most other investigators in recent years.

Luttrell (**1951a, 1955**) noted that there were two types of ascus structure and termed these asci unitunicate and bitunicate. He believed that the bitunicate ascus was limited to and characteristic of the stromatic type of ascocarp and divided the Carpoascomycetes into Loculoascomycetes (Bitunicatae) and Euascomycetes (Unitunicatae).

Recently von Arx and Müller have undertaken the monumental task of studying the type species of all available genera and arranging them according to their ideas of relationship. Their first volume (**1954**) dealt with the Amerosporeae and their second (Müller and von Arx, 1962) with the Didymosporeae. Even though one may not agree with all of their conclusions, these two volumes have brought together an enormous amount of information and will serve as source books for many years to come.

A taxonomic arrangement may serve two functions. It may serve as a convenient means of determination of genera and species. In its extreme form this sort of a system is merely an arbitrary, handy index. The usual keys to genera and species and the numerical listing of characters are examples of this type. The second objective is to show phylogenetic realtionships, which is the final aim of a natural system. Most arrangements, however, are a combination of these two viewpoints. Phylogenetic relationships are often debatable and a matter of personal opinion. The literature is full of transfers of species and genera from one taxon to another, based on personal opinion. In recent years a number of monographic studies have delimited closely related groups of family or ordinal rank. Such taxa will be recognized in the following account. Inasmuch as evolution takes place in a bush-like series of divergent and parallel lines of development, and inasmuch as our present knowledge is fragmentary and limited, our larger taxa are bound to be somewhat arbitrary and artificial. Therefore, it seems less confusing, from a pedagogical point of view, to base these larger taxa on characters such as similarity of morphology and ecologic habit without presupposing any fixed phylogenetic relationship. This is not an innovation, for it has been used in many previous systems.

There have been literally thousands of genera described in the Ascomycetes. It is beyond the scope of this work to consider all of these genera or to attempt to unravel the complicated nomenclature and synonymy. A few examples of genera of the major taxa, those which are best known and most characteristic, will be considered in detail.

Reproduction

Most Ascomycetes reproduce by both asexual spores (conidia) and sexual spores (ascospores). Conidia are pinched off from the tips of vegetative hyphae, which are usually modified and specialized as "conidiophores". These conidiophores form a great variety of branched, simple or whorled hyphae. The conidia may be produced on or in stromata or in enclosed fruiting bodies known as pycnidia. In many species there are two distinct types of conidia, the macroconidium and the microconidium. The macroconidium is relatively large, one to many-celled, with considerable cytoplasm around the nucleus, and each cell can germinate to form a hypha. The microconidium is smaller, has a large nucleus and little cytoplasm, and seldom germinates to form a mycelium. The latter were called spermatia by earlier workers and were looked upon as nonfunctional male gametes. In a number of cases these spermatia have been found to behave in a similar manner to gametes.

In many species of Ascomycetes no conidial state is known, but this does not mean that one does not exist, for it often occurs only under special environmental conditions. On the other hand, many conidial fungi have not been connected with any ascigerous state. In the past many such links have been suggested by observation only. As a matter of fact, in nature there are often many different fungi growing together in a mixture, and this is not a dependable means of assuming relationship. The only sure way of connecting the conidial and ascigerous states is to obtain both of them in single-spore or pure cultures of the fungus. This has been done in many instances. As the result of the lack of connections in many fungi a separate classification of the conidial states has arisen, known as the Fungi Imperfecti. The species within the genera of this classification are based solely upon a morphological similarity and are not necessarily genetically related to one another or to an ascigerous taxon. For this reason they are known as "form genera".

Both homothallism and heterothallism occur in the Ascomycetes. In heterothallic species there are two types of nuclei differing from one another in an allelomorphic pair of genes ("a & A," or "minus" and "plus"). Individuals of the same species possess only one of these two types of nuclei, either type "a" or type "A" in their mycelium. In order to bring about nuclear pairing and fusion and the formation of a zygote (and consequently meiosis and segretation to form the ascospores) two compatible nuclei (a & A) must be brought together in the same cell. The passage of a nucleus from one individual (A) into the mycelium of a second individual (a) is accomplished by various means.

In homothallic species there is either no differentiation in the sexual genes of the nuclei or both types occur in the same mycelium. Fusion of nuclei can, therefore, occur without the transfer of nuclei of the opposite type from one strain to another.

In the Ascomycetes there may be a morphologic differentiation of gametic hyphae which is not necessarily correlated with the heterothallic differentiation of the nuclei. What has been termed as the female gametic organ is the "archicarp" which may take on a variety of forms in different species. Typically it consists of a more or less coiled septate hypha consisting of basal cells, one or several "ascogenous" cells and a "trichogyne" which is usually elongate and non-septate. A more or less modified hypha, known as an "antheridium," may or may not occur and may or may not fuse with the trichogyne to deliver a nucleus to the ascogenous cell. There is little agreement in the literature in this respect.

In many species the microconidia or "spermatia" may fuse with the trichogyne and bring about fertilization. Whatever the mechanism, nuclei of the same or opposite type are brought together in the ascogenous cells. These ascogenous cells give rise to a more or less profuse system of hyphae, termed the "ascogenous hyphae." This ascogenous system is the only place in the Acomycetes where the association and pairing of nuclei takes place to form a dikaryotic mycelium (one composed of cells with paired nuclei). The remainder of the ascocarp is usually haploid vegetative tissue. This is in contrast to the Basidiomycetes where an extensive dikaryotic mycelium is formed.

In most but not all Ascomycetes the asci arise by the formation of an ascus crook or "crozier" which in many respects resembles the clamp connection found in the Basidiomycetes and suggests that these two large groups are descended from a common ancestor. The apical cell of the ascogenous hyphal branches contains two nuclei. If two individuals were concerned in the fertilization of the ascogenous cell, each nucleus was descended from the nucleus of one of the two parents. This cell curves over in the shape of a crook and the two nuclei undergo mitosis. This results in a row of four haploid nuclei. Septa are now laid down in such a manner that a uninucleate ultimate, a binucleate penultimate and a uninucleate antipenultimate cell are produced. The binucleate cell elongates and becomes the young ascus. The two nuclei derived from the two parents now fuse and undergo meiosis and a mitosis to form the, usually, eight nuclei around which the ascospores are delimited. The two uninucleate cells (ultimate and antipenultimate) may fuse to bring together two paired nuclei and develop as another apical cell.

Mutations also commonly occur in the Ascomycetes. If a mutation occurs in genes affecting morphological or physiological characters, it may give rise to heterokaryosis within the same mycelium. Somatic isolation of such a character in one portion of the mycelium may give rise to a "saltation" or modification from the parental type, which is a common phenomenon in these fungi. Heterokaryosis may also occur by the fusion of related hyphae and the passage of nuclei from one mycelium to the other.

Such mutations affecting the gametic hyphae may give rise to what are known as male, female and neuter strains. A female strain is one which produces archicarps but not microconidia. Such a strain would produce ascocarps only when fertilized by microconidia from another strain. A male strain is one which produced microconidia but no archicarps. Such a male strain cannot produce ascocarps, but can fertilize a female strain which has archicarps. Neuter strains produce neither archicarps nor microconidia. They cannot produce ascocarps nor fertilize other strains.

Related strains of species with various mutations can be crossed (particularly with heterothallic species) to produce hybrids. Since there is only one nuclear fusion and one meiosis in each ascus which gives rise to (usually) eight haploid spores, the segregation of characters can be favorably studied in these fungi. In some species at least, there seems to be only one pair of nuclei from which all others in the ascogenous system are derived. All the asci of such an ascocarp, therefore will have the same genetic components but in different combinations. These characteristics make these fungi favorable organisms for the study of genetics. Many important contributions have been made from fungi, particularly the genus *Neurospora* (see that genus), which has been studied in great detail.

The formation of reproductive bodies (ascospores or conidia) is only the first stage in the propagation of the species. The second stage is the dispersal of these spores to a favorable substratum. There have been many ingenious methods of such dispersal evolved by the fungi. Many of them have been described by Ingold (1965). Mason has recognized two types of conidial spores: dry spores with dry walls, as in *Penicillium*, which are dispersed by air currents, and slime spores which have a mucous covering and adhere to one another and are dispersed by rain, insects or birds. Many pycnidial forms have such spores which are extruded through an ostiolar opening as droplets or threads known as spore horns, as in the chestnut blight fungus, *Endothia parasitica*.

Ascospores are distributed in many different ways. Cleistothecial types on vegetable debris and dung may be dispersed by rain, insects or rodents. Some forms with elaborate appendages may be carried in toto by insects or birds. In ascocarp forms with an ostiole the asci or ascospores are pushed or shot out through the ostiolar canal as a result of growth pressure within the ascocarp. Many forms on dung, e.g. *Sordaria*, have an ostiole which is phototropic and bends toward the light. The spores are shot from the ascus with some force. The spores have a mucous covering which sticks to any object which they hit. In dung fungi this is commonly grass or other vegetation, upon which grazing animals feed. Such spores pass through the alimentary canal without harm, and indeed require the action of gastric acids before they are able to germinate upon the dung of the animal. In *Ascobolus, Podospora* etc., the eight spores of an ascus are shot in a single ball. In many perithecial forms on wood, the spores are also shot out to a short distance and can often be seen as a black dust about the ostiole. The bitunicate ascus is an apparatus in which the internal wall carries the spores out of the rigid outer wall which remains behind in the ostiole. In the operculate Discomycetes an osmotic pressure is built up in the ascus. The ascospores are in a single row and when any trigger stimulus such as a slight jar or temperature change occurs, all the operculae of the mature asci pop open and the ascospores are shot out through the elastic opening in the tips of the asci in such numbers that they may be visible as a cloud.

In the case of many parasites, especially those found on leaves, infection takes place on the living host, but the initiation of fruiting bodies does not occur until the close of the growing season. Such ascocarp initials carry the fungus over the winter. The young asci enter a resting state after nuclear fusion, and meiosis and spore formation occurs during the early spring. The ascospores are then distributed by air currents and are carried to the soft tissues of opening buds which they infect. Many such fungi have a conidial state which serves to propagate the fungus during the growing season. *Venturia inaequalis* and species of *Mycosphaerella* are examples of this type.

Some underground fungi, like the Tuberales, have no active means of spore dispersal and the spores are probably carried by burrowing insects, earth worms, rodents etc. Fruit bodies of *Elaphomyces* are often found in the crotches of trees, where squirrels have gnawed and left them.

Thaxter noted that in the Laboulbeniales the ascocarps are often limited to certain parts of the insect's anatomy. This may be due to the method of copulation, during which the spores are transferred to certain favored areas.

CLASSIFICATION

No single classification will satisfy everyone. From what has been said in the historical section, it is obvious that there have been many systems based upon different sets of characters. The overall relationships within the Ascomycetes are fairly clear but the details of evolutionary series are largely a matter of opinion. The earlier, what we may call the "pre-Nannfeldt" arrangements, were largely arbitrary and were based upon the gross morphology of the ascocarp, asci and ascospores. In later years, what may be called the "post-Nannfeldt" era, attempts were made to erect a more natural classification, based upon developmental characters, which would indicate genetic relationships of the organisms concerned. Any resulting arrangement is usually a compromise between the practical and the theoretical. There are sure to be changes with time; our knowledge of the relationships of the thousands of species is never complete and interpretations will always differ.

In order to have a basis for comparison of the changes from the pre-Nannfeldt to the post-Nannfeldt systems, the reader is referred to Luttrell (**1951a**), who has outlined the systems of Lindau (**1897**) and other pre-Nannfeldt writers.

The earlier arrangements served their purpose as a convenient filing system for the determination of genera and species, for they were based on comparatively definite and easily determinable characters. In later years the use of developmental characters of the ascocarp resulted in the formation of several outlines of classification by Nannfeldt (**1932**), Miller (**1949**), Luttrell (**1951a, 1955**), von Arx and Müller (**1954**), and Müller and von Arx (**1962**). None of these, however, has been complete in detail. The following key, which is used in this work, is an attempt to correlate these viewpoints into a general system.

KEY TO ORDERS AND FAMILIES OF ASCOMYCETES

I. Asci borne free on the mycelium; not within a sterile fruit body or ascocarp .... **PROTOASCOMYCETES**
- A. Asci for the most part globose, with 2-8 ascospores ... ENDOMYCETALES (p. 18)
- B. Asci either elongate, clavate and multispored or atypical .. HEMIASCALES (p. 22)
  - 1. Asci elongate, multispored .... Ascoideaceae (p. 22)
  - 2. Asci various, atypical .... Hemiascaceae (p. 23)
  - 3. With both haploid and diploid mycelia .... Spermophthoraceae (p. 24)
- C. Asci clavate, parasitic on green plants .... TAPHRINALES (p. 24)
  - 1. Asci developing from thin-walled ascogenous cells .... Taphrinaceae (p. 24)
  - 2. Asci developing from thick-walled resting spores .... Protomycetaceae (p. 25)

II. Asci borne upon or within a sterile fruit body, or ascocarp ... **CARPOASCOMYCETES**
- A. Asci in an hymenial layer, widely exposed at maturity .... (DISCOMYCETES) (not considered)
- B. Asci within an enclosed fruit body .... PYRENOMYCETES
  - 1. Asci borne in locules within a stroma (wall of stromatic tissue only) (Bitunicatae) .... **LOCULOASCOMYCETES**
    - a. Ascocarp dimidiate to crustose; basal wall incomplete................. .... MICROTHYRIALES (p. 91)
    - b. Ascocarp elongate, opening by a longitudinal slit... HYSTERIALES (p. 97)
    - c. Not as above; ascocarp a globose, pulvinate or effuse stroma
      - (I). Asci borne singly in separate cavities .... MYRIANGIALES (p. 32)
        - (A). Asci globose to saccate, scattered singly in the unaltered stroma tissue..................................... Myriangiaceae (p. 34)
        - (B). Asci clavate to cylindric, separated by elongate elements of the stroma...................................... Dothioraceae (p. 37)
      - (II). Asci borne in fascicles within locules in a stroma
        - (A). Central tissues of the locule unaltered or disintegrated stromal cells .... DOTHIDEALES (p. 40)
          - (1). With a single locule in each stroma
            - (a). Stromata immersed in the substratum.................. .... Mycosphaerellaceae (p. 52)
            - (b). Stromata superficial on the substratum
              - X. Ascostromata walls gelatinous at maturity............ .... Englerulaceae (p. 61)
              - XX. Ascostromata walls not gelatinous .... Perisporopsidaceae (p. 62)
          - (2). With more than 1 locule in a stroma .... Dothideaceae (p. 41)

(B). Central tissues of the locule composed of filiform paraphysoids . . .
. . . . PLEOSPORALES (p. 64)
(1). Ascostromata with flattened ostioles opening by a longidutinal slit . . . . Lophiostomataceae (p. 85)
(2). Not as above; ostioles perforate
(a). Ascostromata superficial upon the substrate
X. Ascospores with germ pores or germ slits . . . .
. . . . Sporormiaceae (p. 89)
XX. Ascospores without germ pores or slits
Y. Ascostromata on superficial torulose or ribbon-like mycelium . . . . Capnodiaceae (p. 59)
YY. Ascostromata not as above
Z. Ascostromata with elongate necks; mostly caespitose . . . . Coryneliaceae (p. 86)
ZZ. Ascostromata not with elongate necks; mostly setose . . . . Dimeriaceae (p. 83)
(b). Ascostromata immersed within the substratum
X. Ascospores 1-celled, brown with sculptured epispore (striate or roughened) . . . . Mesnieraceae (p. 85)
XX. Ascospores greenish-yellow to olive-brown, often apiosporous . . . . Venturiaceae (p. 77)
XXX. Ascospores 4-celled, with central cells shorter, and with a gelatinous envelope . . . . Massariaceae (p. 90)
XXXX. Ascospores various, not as above . . . . Pleosporaceae (p. 65)

2. Asci in true perithecia with a differentiated wall (Unitunicatae) . . . .
. . . . . EUASCOMYCETES
a. Asci irregularly scattered within an enclosed fruit body (cleistothecium) . .
. . . . PLECTASCALES (p. 103)
(I). Fruit body stalked
(A). Asci scattered throughout a globose to pulvinate head . . . .
. . . . Onygenaceae (p. 106)
(B). Columns of asci between sterile veins; open above . . . .
. . . . Trichocomaceae(p. 116)
(II). Fruit body not stalked
(A). Wall of fruit body a loose weft or peridium of modified hyphae. . . .
. . . . Gymnoascaceae (p. 104)
(B). Fruit body with a firm wall
(1). Hypogeous (subterranean) fungi
(a). Core corky or woody; spores hyaline . . . . Mesophellaceae
(b). Core cottony or fleshy; spores brown Elaphomycetaceae (p. 116)
(2.) Not hypogeous
(a). Ascocarp ascohymenial, arising from a coiled archicarp
X. Spores with a small basal appendage-like cell. . . . .
. . . . Tripterosporaceae (p. 114)
XX. Spore not so appendaged . . . . Eurotiaceae (p. 107)
(b). Ascocarp arising as a parenchymatic stroma . . . .
. . . . Phaeotrichaceae (p. 115)

b. Asci in fascicles, usually evanescent, in perithecia with more or less elongate ostioles . . . . MELANOSPORALES (p. 116)
(I). Perithecia covered with spiral or branched hairs or setae . . . .
. . . . Chaetomiaceae (p. 118)
(II). Perithecia smooth or with simple hairs
(A). Ascospores pale straw color to red brown . . . Melanosporaceae (p. 117)
(B). Ascospores hyaline; ostioles very elongate . . . . Ceratocystaceae (p. 120)

c. Asci in a more definite hymenial layer, not evanescent
(I). Asci with a thickened disc or ring in the apex which is often amyloid . . . . XYLARIALES (p. 122)
(A). Ascospores filiform, sigmoid, hyaline. Perithecia often with hairs or setae. . . . . Lasiophaeriaceae (p. 132)
(B). Ascospore ellipsoid, brown
(1). Perithecia superficial, not in a stroma
(a). Ascospores usually with appendages or a mucous sheath; mostly on dung or vegetable remains . . . . Sordariaceae (p. 123)
(b). Ascospores without appendages; mostly on vegetable remains or wood . . . . Roselliniaceae (p. 127)
(2.) Perithecia imbedded in a well developed stroma . . . . Xylariaceae (p. 132)
(C). Ascospores not as above, not filiform nor with germ pores or slits. . . . . Amphisphaeriaceae (p. 179)
(II). Asci not with an amyloid apical disc
(A). Ascospores allantoid and yellow or brown in mass . . . . ALLANTOSPHAERIALES (p. 136)
(1). Perithecia non-ostiolate; asci extruded in a gelatinous mass . . . . Coronophoraceae (p. 141)
(2). Perithecia ostiolate; not as above
(a). Perithecia superficial, with a stalk-like base . . . . Nitschkiaceae (p. 143)
(b). Perithecia immersed; not stalked
X. Perithecia clustered but not in a stroma. . . . . Calosphaeriaceae (p. 142)
XX. Perithecia imbedded in a stroma . . . . Diatrypaceae (p. 137)
(B). Ascospores not allantoid and colored in mass; (allantoid and hyaline in some cases)
(1). Asci with a refractive ring in the apex; asci usually deciduous in the perithecium. DIAPORTHALES (p. 144)
(a). Ascospores allantoid and hyaline. . . . . Valsaceae (p. 159)
(b). Ascospores not allantoid
X. Perithecia scattered or grouped within an entostroma or beneath an ectostroma . . . . Diaporthaceae (p. 151)
XX. Perithecia scattered singly within the host or substratum; not in a stroma. . . . . Gnomoniaceae (p. 145)
(2). Asci not with a refractive ring in the apex nor with deciduous asci
(a). Perithecia with pseudoparaphyses growing downward from above; perithecia fleshy and bright colored . . . . HYPOCREALES (p. 193)
X. Asci with a simple thickened apex; stroma, when present, fleshy . . . . Hypocreaceae (p. 195)
XX. Asci with a thickened apical wall penetrated by a canal; perithecia in or on a loose subiculum . . . . Hypomycetaceae (p. 203)
(b). Perithecia with true paraphyses growing up from the base, or none
X. Ascocarp a minute parasite on the exoskeleton of insects consisting of a receptacle bearing appendages, and a perithecium, and/or antheridia. . . . . LABOULBENIALES (p. 211)

Y. Spermatia formed endogenously from phialid-like antheridia
Z. Antheridia borne singly on branches of the appendages .... Laboulbeniaceae
ZZ. Antheridia compound; several opening into a common chamber .... Peyritschiellaceae
YY. Antheridia lacking; spermatia budding exogenously. .... Ceratomycetaceae
XX. Not as above; typical perithecia with or without a stroma
Y. Asci long cylindric with a thickened apical plug penetrated by a canal; ascospores long filiform; parasitic; perithecia mostly bright colored......... .... CLAVICIPITALES (p. 206)
Z. Parasites of grasses and sedges .... Clavicipitaceae (p. 207)
ZZ. Parasites of insects and other fungi............ .... Cordycipitaceae (p. 210)
YY. Not with the above ascus, perithecial and ascospore characters
Z. Perithecia not ostiolate.... ERYSIPHALES (p. 163)
(x). Superficial mycelium hyaline .... Erysiphaceae (p. 164)
(xx). Superficial mycelium brown Meliolaceae (p. 169)
ZZ. Perithecia ostiolate....... SPHAERIALES (p. 172)
(x). Marine fungi, found in salt walter; ascospores mostly with appendages........... .... Halosphaeriaceae (p. 188)
(xx). Leaf parasites with perithecia immersed in the host or a stroma or under a clypeus .... Polystigmataceae (p. 172)
(xxx). Saprophytes on dung; ascospores large, 1-celled .... Seliniaceae (p. 193)
(xxxx). Various, not as any of the above .... Sphaeriaceae (p. 184)

General Discussion of the Relationships and Major Characteristics of Ascomycetes

The theories upon which the proposed keys have been based are still in a state of flux. Much information is still needed before the proper distribution of genera can be reached. There have been well over fifteen hundred genera described within the pyrenomycetous fungi. According to the rules of nomenclature the type collection of the type species of each genus determines the characteristics governing that genus. Many of the descriptions of genera have been incomplete or incorrect regarding the diagnostic characters. The concept of genera, in the past, has not always been in accord with the type collection. The type collections, themselves, have often been immature, in poor condition, or unavailable. The concept and circumscription of genera also has varied with the opinion of the investigator. Thus, many genera have been placed in synonymy or transferred from one family or order to another. It is beyond the scope of this work to follow the details of such synonymy or transfers. It has become mandatory that a careful study of the type collection be made to determine the characteristics of species and genera. This has been done by the monographers of certain groups, but many genera are still in question. Von Arx and Müller (**1954**) and Müller and von Arx (**1962**) have studied probably the greatest number of generic types in the amerosporous and didymosporous pyrenomycetes. Their placement of these genera has, therefore, been followed in general in the following work. These authors have reduced many genera to synonymy and some of these may be questioned. There still remains a large number of genera whose position or validity is in doubt.

In nearly all classifications there is recognized a primitive group of Ascomycetes in which the asci are borne free upon the mycelium, and no sterile fruit body is formed. These are the Protoascomycetes of the present work. In the Saccharomycetaceae are simple one-celled types with typical cell fusion and ascus formation. In the Endomycetaceae a true mycelium first appears. In the Hemiascales there occur several evolutionary trends which seem to reach dead ends, whereas the Taphrinales have become highly specialized and adapted to a parasitic habit.

The Carpoascomycetes, the second large group of the Ascomycetes includes those species in which the asci arise within a sterile fruiting body or ascocarp. In all past systems this group has been divided into two extra-taxonomic divisions, the Discomycetes with fruit bodies which open widely at maturity (apothecia); and the Pyrenomycetes with enclosed fruit bodies with or without a preformed opening. The present work is not concerned with the Discomycetes.

Nannfeldt (**1932**) recognized two large divisions within the Carpoascomycetes (see that discussion, page 26), the Ascohymeniales and the Ascoloculares.

In such forms as *Byssochlamys*, in which clusters of simple globose asci arise within a loose mass of vegetative hyphae, we can see the transition from types such as the Endomycetaceae to those found in the Gymnoascaceae. The Plectascales, for the most part, have small globose asci scattered throughout the tissue of an enclosed ascocarp with an outer wall which is composed of a loose differentiated hyphal weft in the Gymnoascaceae and a firm compact wall in the Eurotiaceae. The asci of the Plectascales are thin-walled and unitunicate and could give rise to the operculate Discomycetes, which have an ascohymenial development and unitunicate asci, through such genera as *Rhyparobius* or certain of the Ascobolaceae in which the ascocarps are entirely enclosed when young but apothecium-like at maturity.

The Plectascales probably include simple forms of a number of evolutionary lines, as suggested by Cain (**1948, 1956**). Most species are of the ascohymenial type, but some, like *Phaeotrichum*, arise as an ascostroma. The genus *Preussia* has clavate asci and septate brown ascospores, as in *Sporormia*, which has been placed in the Bitunicatae.

The development of clavate to cylindric asci and their orientation into a fascicle or a hymenial layer, is seen to take place in the Melanosporales, where the ostiolar neck also occurs for the first time. The organization of the asci into a fascicle or hymenial layer in the base of the perithecium is common to all the remaining orders of the Euascomycetes. These orders as recently conceived have been based on rather subtle groupings of characters indicating an evolutionary relationship which included a variety of ascocarp types from simple perithecia to well developed stromata.

The Xylariales have ascospores with a single brown cell with germ pores or germ slits in the wall or with gelatinous appendages or envelopes, and asci with ring-like thickenings in the upper wall which stain blue with KI. The order includes simple superficial perithecia on dung or decaying wood and compound, well developed stromata on woody stems.

The Allantosphaeriales are characterized by allantoid ascospores which are colored in mass and include a variety of perithecial forms from simple to stromatic ones, mostly saprophytic upon woody substrata.

The Diaporthales are an order with rather indefinite boundaries, but characterized by asci with a refractive ring in the apex, which is not stained with KI, and which show a tendency to be deciduous and lie at various levels in the perithecium instead of in a definite hymenial layer. They include both saprophytes and weak parasites upon woody and herbaceous stems upon leaves, and both simple and stromatic species.

The Hypocreales and Clavicipitales have perithecia with bright colored and fleshy walls for the most part and again either simple perithecia or stromatic forms. Many species with light colored perithecia have been placed in other orders and these two orders are distinguished on other characters. The Hypocreales have pseudoparaphyses which grow downward from the upper part of the perithecial cavity and the Clavicipitales have long cylindric asci with thickened apical caps and filiform ascospores. The two orders contain parasites on various plants, insects and other fungi.

The Erysiphales include a similar ecological group of superficial leaf parasites with globose perithecia and a superficial mycelium.

The Laboulbeniales are a distinct group of parasites upon the exoskeleton of insects, with a distinctive type of ascocarp.

The Sphaeriales represent a heterogeneous group of three families which do not fit into the conception of any of the other orders of the Euascomycetes. The Polystigmataceae include genera which are leaf parasites that are imbedded in the leaf or in a more or less well developed stroma or beneath an epidermal clypeus. The diagnostic characters of the family are not clear cut and it includes genera closely related to other orders. The Amphisphaeriaceae is based upon species which have asci with a thickened ring in the apical wall staining blue with KI, a character similar to that in the Xylariales. The perithecia are immersed in leaves or stems. The Sphaeriaceae is a family made up of genera which are mostly superficial saprophytes and of a type which do not agree with any of the previous described families.

The Loculoascomycetes, a group in which the ascocarp arises as a stroma, within which the asci are bitunicate and borne in locules, has been recognized by most modern authors. There still seems to be some sorting necessary in the distinctions of the Loculoascomycetes and Euascomycetes. Some taxa include species with both the ascohymenial and the ascolocular type of development. Some species of *Mycosphaerella* have a definitely stromatic origin, whereas others originate as a coiled archicarp. The closely related genus *Didymella* has been looked upon as ascohymenial. The Plectascales are mostly ascohymenial in origin but *Phaeotrichum* is definitely stromatic. Both ascohymenial and ascostromatic types have been described in the Microthyriales. Von Arx and Müller (**1954**) and Müller and von Arx (**1962**), on the basis of the one- and two-celled ascospore types, have recognized that the details of the origin of ascocarp formation are not known in many genera. They have based their Ascoloculares upon the presence of the bitunicate ascus. Here, again, the actual emission of the ascospores has not been seen in many genera. These authors consider any ascus with a thickened double wall as bitunicate. They disagree with the use of the type of the development of the interthecial tissue for the separation of orders, as suggested by Miller and Luttrell. Instead they present the three orders of the Loculoascomycetes as follows:

Myriangiales: Asci globose to saccate, scattered singly in separate locules within the stroma.

Dothiorales: Asci in clusters or hymenial layers within cavities in a stroma. Ascocarps opening by a slit or an irregular rupture instead of a perforate ostiole.

Pseudosphaeriales: Asci in clusters or a hymenial layer within cavities within a stroma. Ascocarp opening by a pore or canal (ostiole).

This arrangement brings together a number of widely separated stromatic types within the Dothiorales. The method of rupture of the ascostroma alone is not sufficient to place all of these in the same order.

The Myriangiales of the system here presented includes those species in which the asci occur singly within separate cavities of the stromatic tissue. In the Myriangiaceae the asci are globose to saccate and are imbedded in the unaltered stromatic matrix. The ascospores are mostly muriform and hyaline. In the Dothioraceae the asci are clavate to cylindric and are arranged in a palisade-like layer and separated by elongated strips of stromatic tissue. The ascocarps of both these families open by a slit or irregular rupture of the overlying tissues.

The flattened or effuse ascocarps as found in such genera as *Johansonia, Phillipsiella, Dothiora*, etc., of the older families Agyriaceae and Phillipsiellaceae, grade off into the inoperculate Discomycetes, some of which, in the Phacidiales, are known to be bitunicate. Other genera, as *Delphinella*, suggest the origin of the Dothideales by the clustering of the asci in a central group.

The Dothideales include species which have their asci arranged in a fascicle or hymenial layer within a differentiated locule in the asco-stroma. The interthecial tissues in the locules of this order are entirely dissolved away by the developing asci or are unaltered or compressed into filiform strips of cells by the crowding of the asci. Transitional conditions of these modifications of locule structure are certain to occur. Species with only one locule have been arbitrarily placed in the Mycosphaerellaceae and those with more than one locule in the Dothideaceae.

In the Pleosporales the central matrix of the locule elongates to form an irregular palisade of filiform cells before the asci arise. The asci push up between these paraphysoids. The genera of Müller and von Arx's Pseudosphaeriales have been included in this order. The details of formation of the interthecial tissues is unknown in many of these genera and these authors have not used this character for the separation of orders. These genera of the Loculoascomycetes need further study and revision.

PROTOASCOMYCETES

The term Protoascomycetes includes those Ascomycetes in which the asci are borne free on the mycelium, or as single cells, but not within a fruit body (ascocarp) of organized sterile tissue.

This group is the logical point of origin of the Ascomycete series. It includes one-celled forms produced in asci, as well as more developed types with true mycelia and ascus formation approaching that found in the ascocarp series.

This division represents a reservoir of evolutionary tendencies giving rise to a number of lines of development, some of which terminate blindly. The simplest forms in this group are the so-called "yeasts" (Endomycetales). Such yeasts are widely distributed in soil, water, sewage, on fruits and many carbohydrate substrata. The term yeast was originally used for all those fungi which reproduced vegetatively by budding. We now know that this includes a very heterogeneous group. Some budding forms are conidial states of higher Ascomycetes, whereas other true Protoascales may form a true mycelium and have various types of gametangia. Some so-called yeasts do not form asci and were known as asporogenous yeasts. These are now placed in the Fungi Imperfecti.

The yeast cell contains a large central vacuole (nucleus, nucleolus, or centrosome), adjacent to which there is a spherical body with a lighter staining center (nuclear vacuole, nucleus) and a peripheral material which takes a chromatin stain (heterochromatin). Within the vacuole there is a variously staining body (nucleus, volutin granule, heterochromatin). Attached to the vacuolar membrane there are granules or threads stretching across it which take a chromatin stain (chromosomes). These structures have been variously interpreted as indicated. Within the cytoplasm there may be mitochondria and glycogen, under certain metabolic conditions. When a bud is formed, a small extrusion is pushed out from some point on the cell surface, and increases in size. The nuclear vacuole puts out an elongated tube which passes into the bud, enlarges at the tip and is cut off as a daughter vacuole. The centrosome body divides and half passes into the bud. The chromosomes also divide.

Guillermond (**1940**) recognized two strains of yeasts which he called "haplobiontic" and "diplobiontic." Haplobiontic yeasts have haploid vegetative cells and produce asci by the fusion of two such cells. Diplobiontic yeasts have diploid vegetative cells and form asci directly from such cells without fusion (these are known as parthenogenic yeasts).

Stelling-Dekker (**1931**) and Lodder and Krieger-van Rij (**1952**) have given us the most comprehensive study of these yeast fungi. A modification of Guillermond's modification of the arrangement of Krieger-van Rij is given below.

ENDOMYCETALES

I. ASCOIDEACEAE: Asci elongate, polysporous, mostly multinucleate.
  A. No basal fusion in ascus formation; multinucleate .... ASCOIDEA
  B. Asci formed by a basal fusion of gametic branches .... DIPODASCUS
II. ENDOMYCETACEAE: Asci not elongate, nor multinucleate.
  A. EREMASCOIDEAE: With a septate mycelium, but no conidia
    1. One genus with gametangial branches .... EREMASCUS
  B. ENDOMYCETOIDEAE: With a septate mycelium, forming arthrospores by fragmentation of the mycelium.
    1. True mycelium with arthrospores and gametangia............. ENDOMYCES
    2. Mycelium rare; division by septation (fission) .... SCHIZOSACCHAROMYCES
  C. SACCHAROMYCETOIDEAE: Budding cells globose or ellipsoid, pseudomycelium (adhering chains of budded cells) or none; budding multilateral.
    1. ENDOMYCOPSEAE: Mycelium present .... ENDOMYCOPSIS
    2. SACCHAROMYCETEAE: Pseudomycelium or none
      a. Not forming a film or pellicle
        i. Spores smooth; always fermentative, conjugation present or absent...
        .... SACCHAROMYCES
        .... (Zygosaccharomyces)
        ii. Conjugation tubes formed but no conjugation
        (a). Spores warty with a ledge.................. SCHWANNIOMYCES
        (b). Spores smooth............................. TORULASPORA
      b. Forming a film or pellicle.
        i. Spore rough, warty, 2 in an ascus................ DEBARYOMYCES
        ii. Spore with a protruding rim, hat-shaped, conjugation present or absent .... HANSENULA
        .... (Zygohansenula)
        iii. Ascospores smooth, reniform, hemispheric or triangular; conjugation present or absent .... PICHIA
        .... (Zygopichia)
    3. NADSONIAE: Cells lemon shaped with bipolar budding; pseudomycelium
      a. Spores round, smooth, conjugating in the ascus ... SACCHAROMYCODES
      b. Conjugation between mother cell and bud; spores warty, brown. NADSONIA
      c. No conjugation, parthenogenic; spores round, hat-shaped ....
      .... HANSENIASPORA
  D. NEMATOSPOROIDEAE: Cells variable; ascospores elongate to filiform
    1. Cells oval; asci 1-spored .... MONOSPORELLA
    2. Cells variable; ascospores needle-like with whip-shaped terminal flagella; 1-8 spores in the ascus in 2 groups .... NEMATOSPORA
    3. Mycelium more definite, asci with 4-32 spores...................... ASHBYA
    4. Spores filiform, without appendages, in a single group .... COCCIDIASCUS
    5. Spores up to 50 in an ascus; shorter, one end pointed; no appendages.......
    .... EREMOTHECIUM
  E. LIPOMYCETOIDEAE: Budding only; asci as protrusions of vegetative cells; spores light reddish, up to 16 in an ascus........................ LIPOMYCES
III. SPOROBOLOMYCETACEAE: Forming ballistospores, discharged from sterigmata; basidiomycetous yeasts
  A. Ballistospores symmetrical .... BULLERA
  B. Ballistospores asymmetrical .... SPOROBOLOMYCES
IV. CRYPTOMYCETACEAE (Torulopsidaceae): Forming conidia only, as blastospores or arthrospores; no asci .... FUNGI IMPERFECTI

In addition to those listed in the outline given above, there are a number of genera with peculiar methods of ascus formation which seem to fit none of the families mentioned. Most of these have been placed in separate families, many of which are monotypic and are sometimes placed in an order Hemiascales, which will be described later. The Ascoideaceae are sometimes placed in this order.

SACCHAROMYCETOIDEAE: This is the sub-family of the typical yeasts, forming a"pseudomycelium" consisting of branching chains of budding cells which have failed to separate, or free budding cells which have failed to separate, or free budding cells. *Endomycopsis* is included because it forms blastospores from its mycelial cells.

Saccharomyces: This is the best known and commonest of the yeasts. It is also of the greatest economic importance. It includes most of the industrially important yeasts. *Saccharomyces cerevisiae* is the brewers' and bakers' yeasts, and its variety *ellipsoideus* is the wine yeast. These yeasts cause alcoholic fermentation, and are used in the production of ethyl alcohol and all sorts of alcoholic liquors from any carbohydrate source. Nearly all yeasts can utilize glucose and dextrose; some can or cannot ferment other simple sugars as maltose, galactose, etc. Amylase and diastase are not produced by yeasts and they, therefore, cannot break down complex polysaccharides such as starch. *Saccharomyces* is also used as a bakers' yeast and as a source of the B vitamins and other side products of their metabolism.

Lindegren (**1949**a) and Winge and his associates (**1935**, etc.) have studied the life cycle and genetics of these yeasts. Winge and Laustsen (**1938**, **1939**) have been able to isolate individual ascospores from the asci of yeasts and to bring about conjugation of known single gametes, and so determine the segregation of genetic characters (Winge and Roberts, **1954**).

Saccharomyces is a diplobiontic yeast and conjugation usually takes place between ascospores while still in the ascus. The buds from such fused ascospores are elongate ellipsoid (long shoot growth) and are diploid. They give rise to asci directly. Such strains were previously called parthenogenic yeasts. Occasionally, however, there may be several ascus daughter bud cells formed before fusion takes place. These cells are globose or shorter ellipsoid (short cell growth) and can fuse to form an ascus. Such strains were formerly placed in a separate genus, *Zygosaccharomyces*. It is now known that both diploid (parthenogenic) cells and haploid (conjugative) ones can be found in the same species. Therefore, the genera *Zygosaccharomyces, Zygopichia* and *Zygohansenula* have been united with their diploid strains under the names *Saccharomyces, Pichia*, and *Hansenula*, respectively. Winge (**1947**),

Winge and Lausten (**1937, 1939**) and Leopold (**1950**) have shown that segregation takes place within the ascus and recombination takes place upon fusion, and that genetic characters are inherited in the yeasts just as they are in higher fungi. The fermentation of particular sugars, for instance, may be controlled by a single gene or several genes (Winge and Roberts, **1950, 1952, 1953, 1955, 1957;** Winge, **1952**). The genes for fermentation of the various sugars are dominant ones.

Schwanniomyces and *Torulaspora*: In these two genera the vegetative cells put forth conjugation tubes, but fail to conjugate, the asci being formed parthenogenetically.

Schizosaccharomyces: In this genus the cells may divide by a true cross-wall, but fall apart to form "arthrospores." In *S. octosporus*, conjugation results in an elongate ascus containing eight spores. Leopold (**1950**) in studying *S. pombe* found that a wild population could contain both homothallic and heterothallic as well as neutral stains, which arose from one another by mutation.

Pichia (*Zygopichia*), *Hansenula* (*Zygohansenula*) nd *Debaryomyces* are the film forming yeasts. Their cells cling together in small chains known as "pseudomycelia" and form a pellicle on the surface of the medium. They are oxidative rather than fermentative and produce esters rather than alcohol. They may give a distinct flavor, as in sherry. *Hansenula* has "hat-shaped" ascospores which are hemispherical with a marginal rim. *Debaryomyces* has rough warty ascospores, whereas those of *Pichia* are smooth.

The NADSONIAE have vegetative cells which are mostly lemon-shaped or apiculate, and the buds are formed from the two poles only. They form pseudomycelia. In *Nadsonia*, when conjugation occurs the nucleus of a newly formed bud migrates back into the mother cell and fuses with its nucleus. A single rough walled ascospore is formed within the mother cell wall. *Saccharomycodes* has round ascospores which conjugate within the ascus. Winge (**1947**) has shown that in *S. ludwigii* there are two pairs of ascospores lying at the two ends of the ascus. Each pair is heterozygous for two factors; normal and abnormal germination and short and long growth. Fusion gives a diploid cell heterozygous for both factors (Winge and Lausten, **1939a**). *Hanseniospora* is found on grapes and is sensitive to small amounts of alcohol. It is diplobiontic but loses its ability to conjugate in culture, and then resembles the form genus *Kloeckera* (of the Fungi Imperfecti).

The genera *Endomyces, Eremascus* and *Endomycopsis* differ from the true yeasts in the formation of a primitive mycelium and specialized gametic branches. *Eremascus* has a mycelium which may branch and become septate by transverse septa into uninucleate cells. In *E. fertilis*,

(Stoppel, **1907**) two neighboring cells put out lateral outgrowths, which grow together and fuse. The nucleus of each copulating cell divides and one passes out into the branch. The branches fuse and swell to form a spherical ascus within which the two nuclei fuse. Meiosis and mitosis then occur and eight ascospores are formed in the swollen tip. In *E. albus* a similar procedure takes place, but the copulating branches coil about one another before forming the swollen ascus, as occurs in many of the higher Ascomycetes. DeLamatar, et al. (**1953**), have described the cytology of *E. albus*. *Endomycopsis* is similar to *Eremascus*, but the cells form blastospores by budding. The asci arise as lateral branches as in *Eremascus*, but these branches do not fuse. Asci may also arise parthenogenetically from vegetative cells. *Endomyces* forms oidium-like arthrospores by the fragmentation of the mycelium. *Endomyces magnusii* has heterogamous copulating branches. One branch becomes clavately swollen to form the "female" branch. It is multinucleate. The other "male" branch is a slender lateral filament with a swollen tip. Apical uninucleate cells are cut off from the tip of each branch, and fuse. The nuclei fuse and undergo meiosis to form four ascospores in the fusion cell.

NEMATOSPOROIDEAE: This subfamily of parasitic yeasts has ascospores which are elongate fusoid to filiform. The vegetative cells are ellipsoid to elongate or lobed and are uninucleate. They may form pseudomycelia.

Nematospora: This occurs as a parasite upon a number of economic plants, causing diseases known as sterigmatomycosis which seem to be transmitted by insects. *Nematospora coryli* is found on the cotyledons of hazelnut, *N. phaseoli* on those of lima beans and *N. lycopersici* on tomato fruits. Any vegetative cell may elongate and give rise to an ascus directly. Eight nuclei are formed in two groups in the ascus. The spores, globose at first, elongate in opposite directions forming two groups of four filiform ascospores. An appendage develops towards the center on each spore. Wingard (**1922, 1925**) states that the ascospores are two-celled with a nucleus only in the lower cells which forms a bud or a germ tube.

Ashbya gossypii is found on cotton bolls and also on coffee, legumes, citrus and tomatoes. It differs from *Nematospora* only in the more definite mycelium with dichotomous branching. It is transferred by hemipterous sucking insects which inject the spores into the host. From four to thirty-two ascospores are formed in two to three bundles in the ascus.

Eremothecium has up to fifty spores in an ascus. It is similar to *Ashbya* but the spores are shorter and broader and curved at the pointed end. The spores lack appendages and are not in bundles. *Eremothecium* is used in the production of riboflavin.

Monosporella is a small genus isolated from the crustacean *Daphne*. The cells form asci without conjugation. The asci have a single, needle-shaped ascospore.

Coccidiascus legeri was isolated from the intestinal tract of *Drosophila*. It has budding oval cells which conjugate to form asci. Eight filiform spores are formed in a single group.

Protascus has thirty-two spores in an ascus.

HEMIASCALES

In this order there are placed a number of genera with atypical types of ascus formation, which do not seem to fit into any previously described order. The asci are free upon the mycelia, however, as in the Protoascomycetes. They probably represent dead ends of evolutionary development in the reservoir of primitive Ascomycetes.

ASCOIDEACEAE: This family has been previously placed in the Endomycetales by various authors. The asci, however, are formed in a different manner than in that order and the family is better included in the Hemiascales. The asci are elongate or clavate to cylindric and contain numerous ascospores. The young asci may be multinucleate and may or may not arise by the conjugation of gametic branches.

Ascoidea rubescens: This species is found in the sap and slime flux of trees. It forms a multinucleate coenocytic mycelium and produces oval conidia by budding. Where the filaments come in contact with the air there are formed terminal, multinucleate, elongate clavate structures which are cut off by a cross wall. These structures were at first considered to be sporangia and the fungus was thought to be related to the Phycomycetes. Varitchak (**1928**) studied its cytology and found that two favored nuclei, of the many in the "ascus," fused and then underwent meiosis and a number of mitoses, forming many haploid "favored" nuclei around which many hat-shaped ascospores were formed. Inasmuch as this is typical of ascus formation, the fungus was placed in the Ascomycetes. The spores are ejected in a mass as a spore horn. After spore discharge, a second "ascus" grows up from the basal wall into the old empty ascus wall to form a second ascus. This again has been compared to the proliferation of sporangia in *Saprolegnia*. Walker (**1931**) saw fusion between the ejected spores, but thought it was a vegetative fusion. If this were a fusion to form a diploid mycelium, it would relate *Ascoidea* to *Spermophthora*.

Dipodascus: This genus has a similar development. In *D. uninucleatus* the mycelium is composed of uninucleate cells, but in *D. albidus* the mycelium is multinucleate (Biggs, **1937**). Oval, one-celled conidia are formed at the tips of the mycelial branches. In each species the asci are formed from gametic protrusions arising from neighboring cells. These outgrowths fuse at their tips and an elongate cylindric tapered ascus grows out from this point of fusion. In *D. uninucleatus* the gametic branches are uninucleate and the young ascus is binucleate, but in *D. albidus* they are multinucleate and the ascus is also multinucleate. In this latter species, there are again two favored nuclei which fuse and then divide to form many haploid nuclei and a polysporous ascus. (This type of ascus formation has been compared to the type of zygospore formation in such Phycomycetes as *Endogone, Basidiobolus* etc., where the zygospore buds out from the fused tip of progametes.) A detailed study of this genus is given by Batra (**1959**).

HEMIASCACEAE: This family name is used for a group of genera which have various anomalous types of ascus formation, which fall in no other family.

Ascocybe: As described by Wells (**1954**), *Ascocybe* has upright conidiophores which branch to form whorls of budding conidia. The asci are borne on "ascophores" which are upright, septate, pigmented hyphae. The tip of the ascophore remains hyaline and gives rise by bud formation to a fascicle of bud-like chains of cells. Beginning at the apex of each chain of buds, these buds are transformed into asci containing four cucullate ascospores with a hat-like rim. Wells follows Zender (**1925, 1925a**) in placing all genera with hat-shaped spores (i.e. *Endomyces, Hansenula, Schwanniomyces, Ascoidea* etc. in a separate family in which *Ascocybe* would fall.

Myriogonium: This was described by Cain (**1948**) from material of *Odontia sudans*. The mycelium, in the hymenium and subhymenium of the *Odontia*, sends up erect branching hyphae. Any two apical cells of such hyphae may fuse, remaining in open connection with the parent hypha. The fusing cells are identical and after fusion swell at one side to form bean-shaped croziers, which may proliferate to form a fascicle of asci in the usual manner. Each ascus contains eight one-celled, hyaline ascospores.

Trichomonascus: This genus was found by Jackson (1947) on *Corticium confluens*. A large ovoid apical cell is formed on an upright hypha. A cap-like cell is cut off from the tip of the ovoid cell (the ascogonium) and is called the trichogyne. A lateral "antheridial" branch grows out from the hyphae bearing the ascogonium and fuses with the trichogyne. The apex of the antheridial cell then grows upward in a spiral manner, and ascospores are delimited in the ascogonial cell.

Helicogonium: This occurs (White, **1942**) in the hymenium of *Corticium* *microsporum*. It has a coiled female branch which fuses with a simple male branch. An ascogonial cell is delimited by a cross wall. This cell increases in size and becomes an ascus, in which eight two-celled ascospores are formed. These ascospores bud within the ascus and the tip of the ascus gelatinizes to release the spores.

The genera *Oscarbrefeldia* and *Conidiascus* (Holterman, **1898**) are two poorly known genera which probably belong in the Hemiascaceae.

SPERMOPHTHORACEAE: The only genus, *Spermophthora*, of this family is unique in the Ascomycetes in having two types of mycelium, one with haploid nuclei and the other with diploid nuclei. This is, in essence, the type of alternation of generations found in higher plants. The haploid mycelium is multinucleate and gives rise to swollen fusoid "sporangia" either terminal or intercalary, which are cut off by cross walls and have a filiform tip or "trichogyne" when terminal. Such sporangia produce numerous fusoid, lunate conidia. The conidia fuse in pairs and germinate to form a mycelium containing diploid nuclei. Such mycelia give rise to ellipsoid "asci" on lateral branches. These asci contain four to eight ascospores which are haploid and give rise, in turn to a haploid mycelium.

TAPHRINALES

This order includes parasites on higher plants which form asci in a more or less thickly scattered manner beneath, or on the epidermis of the host plant. Two families have been recognized.

TAPHRINACEAE: This family includes a single genus, *Taphrina*. Most of the species are heterothallic and the ascospores bud within or outside the ascus forming a yeast-like growth. Before mycelia can be formed and infection take place, there must be a fusion of compatible conidia, in most species. Such a mycelium is "dikaryotic" with paired nuclei and proliferates within the host tissue. Enlarged branching hyphae spread just beneath the epidermis and become septate into differentiated binucleate "ascogenous cells." These cells put out a clavate growth which penetrates the epidermis and becomes the ascus. In some species a septum is formed at the base of the ascus separating it from the basal "stalk cell." Nuclear fusion occurs in the basal cell and a diploid nucleus passes into the ascus where eight ovoid ascospores are formed. These ascospores bud in a yeast-like manner so that the mature ascus is filled with numerous secondary conidia.

Mix (**1949**) has monographed the family and recognizes three prominent host groups: the ferns, on which the asci are widely scattered; the amentiferous tree species; and the tribe Prunea of the Rosaceae. There are three general symptoms produced on the host; a blister or curling of the leaf tissue; a swelling and discoloration of the ovaries; or formation of a witches broom.

The most important species economically is *Taphrina deformans*, which causes a leaf curl of peaches. Mix (**1935**) found that the fusion of conidia is not necessary for infection in this species. The infecting mycelium caused an irregular swelling, and thereby a curling of the leaf tissue. Abundant anthocyanin is formed in the infected tissue giving it a reddish tint. *Taphrina johansonii* occurs in the ovaries of *Populus tremuloides* causing them to enlarge and turn yellow. *Taphrina pruni* occurs on plums and cherries causing a swelling and inflation of the ovaries.

PROTOMYCETACEAE: This family includes three genera, *Protomyces, Taphridium,* and *Mixia*. The life history is similar to that of *Taphrina*, but the "ascogenous cells" are replaced by thick-walled cells known as chlamydospores. In *Protomyces* they are scattered throughout the host tissue; in *Taphridium* they form a subepidermal layer; and in *Mixia* they form within the outer wall of the epidermal cells. The chlamydospores are multinucleate. In *Taphridium* and *Protomyces* they are three-walled. They germinate by the protrusion of the inner wall as a sac-like outgrowth. A large central vacuole is formed and the cytoplasm lines the periphery of the sac. Numerous ascospores are cut out in this peripheral cytoplasm. Such a structure has been called a "synascus" by those placing these species in the Ascomycetes.

Kramer (**1958**), in describing the genus *Mixia* saw it as an intermediate form between the Taphrinaceae and the Protomycetaceae, with some characters reminiscent of the Phycomycetes. The mycelium is coenocytic with few septa as in the Phycomycetes. The chlamydospore is multinucleate and puts out a thin-walled sac with a central vacuole as in the Protomycetaceae, but the peripheral cytoplasm is cut off from the vacuole by a wall or columella as in *Mucor*. Numerous spores are formed in the peripheral cytoplasm and the outer wall of the sac disintegrates, leaving a layer of spores adhering to the columella. Van Buren (**1915, 1922**) has monographed the Protomycetaceae of Switzerland.

CARPOASCOMYCETES

The divisional name Carpoascomycetes is used for those Ascomycetes which form an ascocarp or fruting body in which the asci or hymenium is enclosed more or less within a sterile tissue of vegetative hyphae, in contrast to the Protoascomycetes in which the asci are borne free upon the mycelium.

It is a very large and heterogeneous group. Several thousand genera have been described. It is beyond the scope of this treatise to consider all of these genera or to attempt to unravel the involved nomenclature and synonymy of these genera.

There is still much to be learned concerning phylogenetic relationships within the Carpoascomycetes. Many writers have given their personal interpretation of the arrangement of these fungi, few of which coincide in all details. Nevertheless, a number of monographic studies of smaller taxa have revealed closely related groups of genera and these will be followed. On the other hand, there are a large number of genera which do not fit into any of these groupings and must be placed by an intelligent guess. Such genera have been brought together in the following account into families and orders which show a similarity of morphologic characters or ecologic habitat rather than any known phylogenetic sequence.

It is also necessary for the author to accept the conclusions of many authorities as to diagnostic characters, for it is impossible for any one person to be familiar with all of the numerous character combinations, from first hand.

The Carpoascomycetes are usually divided into two large extra-taxonomic groups, the Discomycetes and the Pyrenomycetes.

The Discomycetes have a fruiting body known as an "apothecium," in which the ascus hymenium is widely exposed at maturity for the distribution of the spores. They are commonly disc-, saucer- or cup-shaped with or without stalks.

The Pyrenomycetes have a fruiting body known as an "ascostroma" or "perithecium," in which the ascus or hymenium is entirely enclosed within a globose or pyriform or tuberculate fruiting body which opens through a pore or "ostiole" or as an irregular rupture of the upper wall. There may be an elongate ostiolar neck through which the spores are ejected.

In recent years the Carpoascomycetes have been divided into two large groups on the basis of the development of the ascocarp. One group, called the Ascoloculares by Nannfeldt and the Bitunicatae or Loculoascomycetes by Luttrell, have a stromatic type of development resulting in an "ascostroma" (pseudothecium or ascoma of some authors). The second group, the Ascohymeniales of Nannfeldt or the Unitunicatae or Euascomycetes of Luttrell, have the so-called "sphaeriaceous" type of development, resulting in a true "perithecium." This emphasis on the detailed developmental characteristics as a major diagnostic character has stimulated many studies of ascocarp development and resulted in an entirely new arrangement of these fungi.

To understand this dichotomy one must understand the distinctions between these two groups. In the Loculoascomycetes, or Bitunicatae, the ascocarp arises by the division, in all directions, of a vegetative cell or a complex of hyphae into a number of more or less isodiametric cells, giving rise to a pseudoparenchymatic stroma. Within this stroma there arise simple binucleate cells or various types of gametic hyphae, or "ascogenous cells" from which the ascogenous hyphal system and the asci arise. The asci as they are formed grow up into, or dissolve their way through this original stromatic matrix. In the more primitive forms each ascus lies in a separate locule. In more advanced types there may be fertile areas delimited within the stroma which contain any number of asci.

According to Luttrell (**1951a, 1955**) the asci of the Loculoascomycetes are all of the "bitunicate" type. Such asci have a double wall. The outer wall is thin and rigid. The inner wall is often thickened, elastic and gelatinous. In water, the outer wall ruptures at some point and the inner wall stretches out beyond the outer one carrying the ascospores with it. This gelatinous wall soon dissolves and frees the ascospores.

In the Euascomycetes, or Unitunicatae, the ascocarp arises as one or sometimes two differentiated gametic hyphae. The so-called female organ is known as an "archicarp" and typically consists of a coiled or twisted hypha which takes a deeper stain and becomes septate. Such an archicarp consists usually of one or several basal cells, one or several enlarged central cells, or "ascogenous cells" and a more or less elongate septate "trichogyne." The archicarp may or may not be accompanied by a male hypha, or "antheridium." There has been a great deal of difference of opinion as to whether such an "antheridium" is or is not present, whether or not it fuses with the trichogyne, and whether or not nuclei pass into the archicarp or whether or not they fuse with archicarp nuclei. At any rate, vegetative hyphae grow out from the basal cells of the archicarp or from neighboring hyphae and envelop the archicarp to form a compact ball of hyphae. These hyphae form the perithecial wall and the perithecial "centrum." The ascogenous cells of the archicarp give rise to a system of "ascogenous hyphae" from which the asci arise. As the asci begin to develop, vegetative hyphae with free ends grow out from the inner wall cells and grow radially upward into the perithecial centrum. The asci grow up between these filaments and often dissolve them. These "paraphyses" apparently serve as a nurse tissue, and may disappear more or less in the mature perithecium. The asci in this group have a single wall layer. They may have various types of thickening of the apical wall and a plug or "apical apparatus" the function of which is still not clear.

This division of the Carpoascomycetes into two large groupings has led to the transfer of many species, genera, families and even orders to new positions. This has stimulated new ideas in the classification of the Ascomycetes which have been followed by succeeding authors. It should be remembered that, for several reasons, any arrangement of taxa, at the present time, should be looked upon as a temporary one which may need adjustment in the future.

1. Comparatively few species have been studied as to the details of their ascocarpic development. As a result many species have been placed merely by extrapolation of our knowledge of these few examples.

2. There is often difficulty in determining the type of development from mature fruiting bodies. In the past a voluminous literature has accumulated attempting to determine the factors characteristic of the two groups. In some species the stromatic character of the wall is easily recognized as it is usually composed of nearly isodiametric, dark-walled cells but this is not always the case. The origin of the interthecial tissues within the perithecial centrum is not so easily determined. There are some five different methods of origin of these interthecial tissues (see Loculoascomycetes and Euascomycetes). These elements often break down in the mature fruit body, however, and a confusing terminology has arisen as a result. This is reviewed by Luttrell (**1965**) who recognizes three types of interthecial tissues in the Loculoascomycetes and two in the Euascomycetes.

3. The correlation of the bitunicate ascus with the stromatic type of development has been largely used to determine this type of fruit body. However, many bitunicate asci are apparently thin-walled, and all asci with thickened walls are not necessarily bitunicate. The only positive identification is by observation of the extrusion of the ascospores.

4. There are many variations within the two types of development which need recognition. There are also intermediate types and often different types of development within apparently closely related species. Luttrell (**1951a**) recognized eight types of ascocarp development.

It may be well, at this point, to define the terms of the interthecial tissues as they will be used in this work.

"Interthecial parenchyma" is the unaltered ground tissue of the original stroma, which may be compressed by the enlarging asci into strips of elongate cells.

"Paraphysoids" are filamentous elements arising by the intercalary elongation of the cental cells of the original stroma, *before* the origin of the asci. They are attached both above and below.

"Pseudoparaphyses" are filamentous elements which arise from the cells of the upper portion of the stromal cavity and grow downwards toward the developing asci, forming a cavity into which the asci grow, between these filaments.

"Paraphyses" are filamentous elements or sterile hyphae, which grow out from the inner wall cells and radiate towards the ostiole, with free ends above.

LOCULOASCOMYCETES

(Ascoloculares, Bitunicatae)

These are the Ascomycetes in which the ascocarp has a vegetative, stromatic, origin. Luttrell (**1955**) believes that this type of development is correlated with the bitunicate ascus. Such asci have a double wall. The outer wall is thin and rigid. The inner wall is usually thickened, elastic and gelatinous. Before release of the ascospores, in water, the outer wall ruptures at some point and the inner wall elongates carrying the ascospores with it.

In the Loculoascomycetes the ascocarp originates by the division of a single cell, a group of cells, or a knot of intertwined hyphae. These cells divide to form a characteristic stroma of isodiametric cells, often arranged in parallel rows. The outer rows become thick-walled and pigmented. The inner cells remain thinner-walled and often hyaline. This type of stroma is characteristic of most species of the Loculoascomycetes and indicates a stromatic development. In other cases, as in the Pleosporales, it is difficult to determine from the mature fruit body whether it arose as a stroma or from a coiled archicarp. In the Loculoascomycetes ascogenous cells or archicarp-like chains of cells are formed within this stroma, rather than as free coils on vegetative hyphae, as found in the Euascomycetes. The type and manner of development of the interthecial tissues has been used to separate genera, families and even orders. The definition of the terms used in this work can be found under the Carpoascomycetes (page 28). There are many intergradations and variations of these types of development. The following tendencies in the evolution of these tissues may be recognized.

In *Elsinoe* and *Myriangium* of the Myriangiaceae, the asci are globose or saccate and are scattered within the stroma parenchyma, which is not altered by their growth. In the family Dothioraceae of the Myriangiales the asci tend to be more clavate to cylindric and stand parallel in an extended layer. Where such asci are numerous and crowded they tend to flatten the intervening stroma parenchyma to form interthecial strips of parenchyma attached above and below. The formation of a flattened, extended layer of hymenial asci grades off from such forms into the Discomycetes. In other genera, as *Delphinella*, the stroma takes on a globose to pulvinate form and the asci form a cluster in the base of the stroma. This type grades off into the unilocular types of the Dothideales. If the asci are few, their pressure, as they develop, forms similar interthecial strips as in *Wettsteinina*. If the asci are numerous, they tend to dissolve away the original tissue of the locules. The other tendency in the Dothideales is the formation of locules in a stroma. This consists of the differentiation of fertile areas of cells within the stroma, within which the ascogenous system and the asci arise. It is not always clear, in unilocular forms whether or not a true locule has been formed. In the multilocular species this locule differentiation is more clear. Within such locules the asci tend to be formed in a fascicle from a clump of ascogenous hyphae in the base of the locule. In the Pleosporales another variation occurs. In *Pleospora* and *Leptosphaeria*, for instance, the tissues in the center of the locule undergo an intercalary elongation to form filiform strips of cells *before* the asci arise. The asci grow up between these "paraphysoids," which may or may not be dissolved in the process. In the Microthyriales the stroma originates as a more or less radiate shield beneath which a stromatic layer originates. Luttrell (**1944, 1946**) has demonstrated that in some species, the interthecial tissues originate as a downgrowth of hyphae from the cells just beneath the shield. These would fall under our definition of pseudoparaphyses. Luttrell (**1951a, 1955**) has reviewed these types of development and suggests the following three types within the Loculoascomycetes:

1. The *Pseudosphaeria* type of von Höhnel, in which the original stromatic centrum tissue remains unaltered and the asci arise singly within separate locules within the stromatic matrix, i.e., *Elsinoe, Dothiora*.

2. The *Dothidea* type, in which the stroma contains one or more differentiated locules of pseudoparenchymatous tissue, in which a fascicle of asci arise and expand by the dissolution of this locule parenchyma, i.e., *Dothidea, Botryosphaeria.*

3. The *Pleospora* type, in which the stroma contains one or more such differentiated locules, but in which the central cells elongate to form a parallel mass of filaments, before the asci arise, i.e., *Pleospora, Leptosphaeria.* The asci push their way up between these filaments and dissolve them to a more or lesser extent.

In the present account the centrum of the *Pseudosphaeria* and *Dothidea* types is referred to as interthecial parenchyma, that of the *Pleospora* type as "paraphysoids." In the Microthyriales and *Sporormia* elements have been described as downgrowing "pseudoparaphyses."

Von Arx and Müller (**1954**) and Müller and von Arx (**1962**) recognize three orders in the Loculoascomycetes, as follows:

I. Asci in separate locules, separated by tissues of the stromatic matrix. MYRIANGIALES
II. Asci grouped in differentiated locules
  A. Ascocarps opening by a slit or irregular rupture................ DOTHIORALES
  B. Ascocarps opening by a punctate ostiole or ostiolar neck . PSEUDOSPHAERIALES

The use of the rupture of the ascostroma for dispersal, for ordinal separation, results in the association of widely different, and to me, unrelated taxa. Von Arx and Müller's Dothiorales, for instance, includes such widely divergent families as the Atichiaceae, Botryosphaeriaceae, Asterinaceae, Arthoniaceae, Hysteriaceae and Patellariaceae. *Botryosphaeria,* as they describe and figure it, on the other hand, has definite ositioles, and the Micropeltaceae, although related to the Asterinaceae, are placed in a separate order.

The order Dothideales (Lindau, in Engler & Prantl, **1897**) was erected for species with stromatic fruit bodies. The order Myriangiales was erected by Starback (**1899**) for those stromatic forms in which the asci were borne in locules separated by tissue elements of the stroma (pseudoparenchyma). In 1907, von Höhnel pointed out that certain species formerly placed in the Sphaeriales actually had a stromatic type of development, and erected the family Pseudosphaeriaceae for them.

Theissen and Sydow, in their monograph of the Dothideales (**1917**), limited the members of this order to those species which had more than one locule in a stroma. Species with only one locule in the stroma (resembling perithecia) were placed in the Pseudosphaeriales. Miller (**1928, 1949**) objected to the separation of these two orders on the basis of the number of locules in the stroma, because of many intermediate forms. In the Dothideales, as he proposed, the central tissue of the fertile locule remains parenchymatic and the asci push or dissolve their way through it. In the Pseudosphaeriales, on the other hand, the asci grow up between filamentous elements of the stroma, which he believed grew downward from the upper portion of the wall. Luttrell (**1951a, 1955**) agreed with this separation. The so-called "paraphysoids" were shown, however, to arise by the elongation of the central elements or cells of the fertile locule (Wehmeyer, **1955**). Luttrell also pointed out that the type genus of von Höhnel's Pseudosphaeriaceae, *Wettsteinina*, had a dothideaceous type of development according to this viewpoint. He, therefore, adopted a new ordinal name, Pleosporales, for the group of taxa included under Pseudosphaeriales.

Nannfeldt, Miller and Luttrell all place the Microthyriales (Hemispheriales) in the Loculoascomycetes because of the stromatic origin of the ascocarp and the thick double-walled asci. Some members of this group have a parenchymatic ground tissue and others have filiform paraphysoids or pseudoparaphyses, but they will all be considered here under this same order, the Microthyriales, because of the characteristic ascocarp and common ecologic habitat.

The Hysteriales have been shown to have a stromatic development by Luttrell (**1953**), but differ in the elongate fruit body, elongate slit-like opening and ecologic habitat. These orders may be keyed out as follows:

I. Ascocarp flattened, shield-shaped or crustose, a thyrothecium, wall often radiate . . . . MICROTHYRIALES
II. Ascocarp elongate or vertically flattened, opening by an elongate slit . . HYSTERIALES
III. Ascocarp globose, pulvinate or effuse, immersed, erumpent or superficial
   A. Asci borne in separate locules separated by parenchyma or other modified stromatic tissue .... MYRIANGIALES
   B. Asci borne in a fascicle of a few or many in a differentiated area of the stroma
      1. Central tissue of the locules remaining parenchymatic or modified by the developing asci .... DOTHIDEALES
      2. Central tissue of the locules elongating to form paraphysoids or pseudoparaphyses before the formation of the asci................ PLEOSPORALES

From the previous discussion, it can be seen that various investigators have used different sets of characters for the separation of families and orders. No one of these is infallible. The arrangement used in the following pages follows, in general, that suggested by Luttrell (**1955**). However, Luttrell is vague as to the placement of genera and families. Very few of these genera have been studied in detail to determine the origin of the interthecial tissues of the locule and it is difficult to interpret the origin from the mature ascocarp. Both more or less unaltered parenchyma (i.e. *Mycosphaerella*) and interthecial filaments (i.e. *Didymella*) can be found in the same family. The term paraphysoids has been generally applied in the literature to both interthecial parenchymatic strips formed by compression of the developing asci and to intercalary elongation of locule parenchyma before the formation of asci (paraphysoids as used in the present work). As a result many genera and even some families must be placed in a more or less arbitrary manner.

MYRIANGIALES

The members of this order have their asci borne in separate locules within the stromatic matrix of the ascocarp. The species are largely tropical or southern in their distribution. They occur as parasites or saprophytes on higher plants or as hyperparasites on plant lice or on other fungi. The stromata are mostly crustose to pulvinate and have a poorly developed upper wall which ruptures irregularly and not with a definite ostiole. The bitunicate asci are globose to saccate or broad ellipsoid in one family and clavate to cylindric in the other. The spores are mostly septate, either phragmosporous or muriform.

Many of the genera are poorly known and opinions differ as to which family or order they belong. The following is an outline of the many families which have been placed in this order.

I. Asci globose to saccate, separately scattered in the unaltered ground parenchyma of the ascocarp.

Myriangiaceae (*Elsinoeae, Plectodiscellaceae*): Stroma fleshy, asci scattered in a polystichous manner (i.e. in more than one layer).

Myxomyriangiaceae: Stroma with an outer sclerotial, gelatinous layer.

Saccardiaceae: Asci monostichous, in a single row.

II. Asci clavate to cylindric, mostly separated by elongate or filamentous elements of the stromatic matrix.

Atichiaceae: Stroma gelatinous on a basal conidial stroma.

Schizothyriaceae: (of von Arx and Müller). A mixture of dimidiate, and pulvinate superficial forms including genera of the *Cookellaceae, Argyrieae* and *Phillipsiellaceae*.

Arthoniaceae: Cryptotheciaceae: Erumpent forms often associated with algae (lichenized).

Dothioraceae: Pulvinate forms placed mostly in the Dothiorales by von Arx and Müller.

Stephanothecaceae (*Microthyriellaceae*): A family proposed by Luttrell with dimidiate "thyrothecia" of the Microthyriales, but with asci scattered separately in a parenchymatic tissue; later made a synonym of *Stephanothecaceae* (Luttrell, **1955**).

Von Arx and Müller included all the genera with globose to saccate asci in their order Myriangiales. They recognized the families Atichiaceae, Arthoniaceae and Schizothyriaceae in their order Dothiorales. These are brought together in this account as two families:

I. Asci globose to saccate, scattered separately throughout the stroma; arranged polystichous or monostichous, separated by the unaltered parenchyma cells of the stromatic matrix .... MYRIANGIACEAE

II. Asci clavate to cylindric, separated by elongated elements of the stromatic matrix and lying more or less parallel in a flat monostichous layer .... DOTHIORACEAE

All of these fungi constitute a reservoir of primitive types from which various lines of the stromatic Ascomycetes (Loculoascomycetes) have been developed. As a result there are representative primitive types which grade off into other orders. The grouping of clavate asci into fertile locules gives rise to simple unilocular forms as are found in *Leptosphaerulina* and *Wettsteinina*. In *Schizothyrium* and similar members with dimidiate ascocarps are suggested the origin of the Microthyriales with asci in separate locules. Species of *Phillipsiella, Plochmopeltis* and similar genera provide a gradation into the Discomycetes, with a flat hymenium of cylindric asci and the development of an epithecium from paraphysate tips. *Dothiora, Bagnisiella,* etc., could be the origin of pulvinate members of the Dothideales.

Inasmuch as there is no completely satisfactory arrangement of these genera of the Myriangiales, a general discussion of generic types must suffice.

MYRIANGIACEAE:

The family, as outlined here, brings together all those stromatic Ascomycetes in which the asci are globose to saccate and located in individual locules within the unaltered stromatic parenchyma. The asci may be monostichous (Saccardiaceae) or polystichous (Myriangiaceae).

Von Arx's (**1963**) monograph of the order Myriangiales corresponds very closely to this concept of the family Myriangiaceae, and the following is an adaptation from his key to the genera.

I. Hyperparasites on other fungi
  A. Ascospores dark brown at maturity . . . . COOKELLA
  B. Ascospores hyaline or rust red
    1. Ascostroma discoid or pulvinate . . . . ULEOMYCES
    2. Ascostroma crustose, often with radiate margin . . . . PYCNODERMA
II. Not hyperparasites; parasites or saprophytes
  A. Ascostroma borne on a superficial mycelium
    1. Ascospores 2-celled, mycelium brown . . . . ALLOSOMA
    2. Ascospores more than 2-celled
      a. Ascospores brown when mature . . . . DICTYONELLA
      b. Ascospores remaining hyaline or yellowish
        i. Superficial mycelium brown, dematioid . . . . BYSSOGENE
        ii. Superficial mycelium hyaline or light-colored . . . . SACCARDIA
  B. No superficial mycelium bearing the ascostromata
    1. Ascostromata on leaf hairs or glands
      a. Ascospores spherical 1-celled . . . . HYALOTHELES
      b. Ascospores septate, elongate
        i. Ascospores 2-celled . . . . MICULARIA
        ii. Ascospores more than 2-celled
          (a). Asci many spored . . . . XENODIUM

(b). Asci 4- to 8-spored
(1). Ascostromata globose to tuberculate without a basal stratum. .... MOLLERIELLA
(2). Ascostromata crustose to pulvinate, borne on a radial basal stroma.................................. SACCARDINULA
2. Ascostromata not on leaf hairs or glands
a. Ascostromata immersed or erumpent
i. Ascostromata small, crust-like........................... ELSINOE
ii. Parasitic in an immersed fungus stroma; ascostromata light colored.. .... ULEOMYCES
b. Ascostroma superficial or soon erumpent
i. Ascostroma small, flat, with a radiate shield
(a). Asci in a ring about a sterile center .... STEPHANOTHECA
(b). Asci not so oriented .... PYCNODERMINA
ii. Ascostromata pulvinate, discoid, or globose without a radiate shield
(a). Ascostromata spheric, anchored by a foot into the stromata; asci spheric, polystichous............................ DIPLOTHECA
(b). Ascostromata pulvinate or discoid; no stomatal hypostroma
(1). Ascostromata basally attached throughout.... BITANCOURTIA
(2). Ascostromata with a sterile central hypostroma in the substrate; or superficial
(a). Asci spheric, polystichous in a fertile stroma on a sterile base................................... MYRIANGIUM
(b). Asci broad clavate in 1 or 2 layers
(1). Ascostromata small, light colored; ascospores 2-celled . ... BUTLERIA
(2). Ascospores many-celled; ascostromata dark colored
(a). Ascostroma erumpent, discoid, pulvinate; leaf parasites................................ ANHELLIA
(b). Ascostroma on a large sterile stroma; asci clavate with paraphysoids...................... ANGATIA

Elsinoe: The genus *Elsinoe* includes a large number of species, many of which have been described by Jenkins and Bitancourt. The genera *Plectodiscella* and *Isotexis* are usually considered as synonyms. The species of *Elsinoe* occur as light crust-like stromata parasitic on the surface of leaves, stems and fruits of higher plants. The ascostromata are formed as light colored pseudoparenchyma within the host tissue and become erumpent through the epidermis. They are somewhat indeterminate in shape but with a slightly differentiated surface layer. The asci are globose to saccate, bitunicate and scattered within individual locules throughout the stroma. The spores are hyaline or slightly colored, usually three-septate and often with a longitudinal septum in one or two central cells. The conidial state is of the form genus *Sphaceloma*. The conidiophores are short, one-celled or one-septate, dark brown and formed in tight bundles or layers on the surface of the young stroma. The conidia are hyaline, sometimes turning yellow or pale brown, ovoid to ellipsoid, one-celled, occasionally cylindric. Small globose microconidia have also been reported for some species.

Many species cause diseases of higher plants, known as leaf spots, anthracnose or scab. *Elsinoe rosarum* causes an anthracnose of roses (Jenkins, **1938**; Jenkins and Bitancourt, **1957**). *Elsinoe piri* causes fruit spot of apples and pears (Jenkins, **1932**). *Elsinoe fawcetii* causes a scab of sour orange (Jenkins, **1931b**) and *Elsinoe australis* a scab of sweet orange fruit (Bitancourt and Jenkins, **1936, 1937**). *Elsinoe canavaliae* causes a scab of lima beans and other legumes (Jenkins, **1931, 1931b**).

Bitancourtia is very similar to *Elsinoe* but is more superficial and dark colored.

Uleomyces is also similar to *Elsinoe* but von Arx (**1963**) limits it to hyperparasites upon other fungi. *Cookella*, with brown ascospores and *Pycnoderma* with shield-shaped ascostromata are also hyperparasites.

Myriangium: The species of this genus are parasitic on plant lice on the branches of woody plants. A perennial sterile stroma is formed over the plant louse and fertile stromata arise from this stroma in the fall. The fertile stromata are discoid and resemble apothecia, but the asci are globose to saccate and scattered throughout the fertile stroma in individual locules. Miller (**1940**) studied American species.

Miller (**1938**) has described the development in *M. duriaei* and *M. curtisii*. The archicarp arises beneath the periphery of a fertile stroma. It consists of an upright, coiled, septate hypha containing an attenuated apex, the trichogyne, several enlarged central cells, the ascogenous cells, and several stalk cells. Miller saw a filamentous hyphae alongside the archicarp which might be considered an antherifium, but no fusion took place with the archicarp. No spermatia are formed. The primary ascogenous cell divides repeatedly to form a chain of deep staining, multinucleate cells. Vertical divisions give rise to lateral ascogenous hyphae which are binucleate. Terminal cells of these hyphae enlarge and then the pair of nuclei divide, and the cells divide vertically to form two binucleate cells. One of these binucleate cells enlarge and the two nuclei divide to form a row of four nuclei. Two cell walls are then laid down to cut off a central binucleate cell and two end cells which are uninucleate. This procedure apparently corresponds to the formation of croziers in other Ascomycetes. The binucleate cell enlarges and its two nuclei fuse and it becomes an ascus. Meiosis occurs in December or January and the ascospores mature in March or April. The ascospores are muriform and hyaline. The asci lie in pockets of the surrounding ascogenous hyphal cells which make up the fertile stroma.

The genera *Allosoma, Dictyonella, Byssogene*, and *Saccardia* have their ascostromata formed on a superficial mycelium.

The genus *Piedraia* has been placed in a separate family, the Piedraiaceae, under the Myriangiales. Its ascostromata are found on human hair. The dark, hard ascostromata arise under the cuticle through which they are erumpent. The asci occur singly or in groups in individual locules within a pseudoparenchymatic stroma. The ascospores are fusoid, curved, hyaline, one-celled and tapered at the ends.

DOTHIORACEAE:

The asci are clavate to cylindric and arranged in a single layer lying parallel to one another as in a hymenium, but separated by elongate elements of the stromatic matrix. These stromatic elements differ from paraphysoids in being formed by the compression of cells of the ascostroma by the developing asci, whereas paraphysoids are elongated elements which arise before the asci are formed. The family is difficult to circumscribe for there are many genera which lie on the borderline between the Discomycetes, the Dothideales, the Pseudosphaeriales and the Microthyriales.

The following is a key to some of these genera.

I. Spores 1-celled
- A. Ascostroma attached by a broad base.......................... BAGNISIELLA
- B. Ascostroma attached by a central foot .... YOSHINAGAIA

II. Spores 2-celled
- A. Ascostroma pulvinate, grouped on a basal stroma, asci cylindric, many-spored .. .. DELPHINELLA
- B. Ascostroma disc- or shield-shaped, or crustose, asci cylindric in flat layer, 8-spored (or less)
  - 1. Ascostromata or mycelium with setae
    - a. Ascostromata flat crust-like .... CHAETOPLEA
    - b. Ascostroma disc-like or pulvinate
      - i. Ascostroma with setae, ascospores brown......... CHAETOSCUTULA
      - ii. Ascostroma tomentose, ascospores hyaline............. JOHANSONIA
  - 2. Ascostromata and mycelium without setae
    - a. Ascostroma disc-shaped or pulvinate; sharply outlined
      - i. Ascostroma with a basal stalk, on a basal mycelium..... PSEUDODISCUS
      - ii. Ascostroma sessile on the substrate.................... PHILLIPSIELLA
    - b. Ascostroma flat, crust- or shield-shaped
      - i. Ascostroma with branched paraphyses............... PLOCHMOPELTIS
      - ii. Ascostromatic filaments not branched
        - (a). On a basal parenchymatic stroma; mycelial layer slimy............. .... HENNINGSIELLA
        - (b). Ascostroma shield-shaped; basal layer poorly developed; with mycelial margin
          - (i). Ascostroma brown to black .... SCHIZOTHYRIUM
          - (ii). Ascostroma hyaline to gray...................... LEPTOPHYMA

III. Spores with many cells; phragmosporous
- A. Asci 8-spored .... LEPTODOTHIORA

B. Asci many-spored
1. Hyperparasites on other fungi .... ENDODOTHIORA
2. Not on other fungi .... SYDOWIA
IV. Spores muriform
A. Asci 8-spored .... DOTHIORA
B. Asci many-spored .... KEISSLERINA

Schizothyrium and *Leptophyma* have shield-shaped ascostromata and are considered under the Microthyriales.

Phillipsiella has a flat, disc-shaped fruit body with a hymenium as in the Discomycetes, but has an upper parenchymatic wall.

In *Plochmopeltis* this upper covering is an incipient "epithecium" of branched paraphyses as is found in the Patellariaceae.

Müller (1963) described a genus *Epibelonium*, in his family Schizothyriaceae which would be placed in the Dothiorales. The young ascocarps are enclosed but open up widely to appear as a short stalked apothecium. The asci are thin walled, cylindric, but bitunicate (?). He figures paraphysoids (?) between the asci which form an epithecium above the asci. This is another intermediate form between the Dothioraceae and the Discomycetes.

Luttrell (1960) described the development of the ascostroma in *Dothiora schizospora*, on *Symphoricarpos orbiculatus*. The ascostromata first appeared in June as small "pimples" on the surface, caused by the development of a pulvinate mass of hyaline parenchyma beneath the periderm. The outer cell layer of this stroma became thick-walled and pigmented. Microconidial locules were formed in the upper layers either of separate stromata or in the ascostromata themselves. The microconidal locules were formed by the fragmentation of the protoplasts of "spore mother cells" (as in *Dothidea*). These cells budded off hyaline, ellipsoidal microconidia. Ascogonia with trichogynes were formed in the base of ascostromata. The trichogynes extended above the surface of the ascostroma where they formed bulbous tips. Microconidia were seen lying against such bulbous tips but no definite fusion or transfer of nuclei was observed. In the central area of the ascostroma a broad, flat, disc-shaped layer of very small parenchyma cells was differentiated, in contrast to the more spherical locules found in the Dothideales. At this time the ascogonial cells of the archicarp became divided into binucleate cells which in turn gave rise to ascogenous hyphae, which formed a thick layer in the base of the differentiated locule region. Numerous asci arose directly from these ascogenous hyphae, without the intervention of croziers, to form a parallel layer or hymenium of asci. These asci dissolved away the intervening locule parenchyma. This development is intermediate between that of *Dothidea* and that of certain bitunicate Discomycetes.

The genera *Bagnisiella* and *Delphinella* (*Rehmiellopsis*) represent transitional forms. *Bagnisiella* has parallel standing asci separated by columns of elongated parenchyma and represents the source of the *Dothiora* - Discomycetes line. It has two-celled ascospores. *Delphinella*, on the other hand, has pulvinate ascostromata with a fascicle of asci in a locule, but with the asci separated by strips of elongated locule parenchyma, so that the asci lie in separate cavities. It is the source of the uniloculate Dothideales.

Rehmiellopsis was given as a synonym of *Delphinella* by Müller and von Arx (**1962**). It has also been considered as a polysporous form of *Mycosphaerella. Rehmiellopsis abietis* (*R. bohemica*), in Europe and *R. balsameae*, in America, are the cause of a blight of the needles of fir. These species have thick-walled, globose ascostromata without conspicuous ostiole, and polysporous asci. *Rehmiellopsis abietis* was first described as *Sphaerella* (*Mycospherella abietis*). Bubak placed it in *Rehmiellopsis* because of the polysporous asci. Waterman (**1945**) and others consider *Rehmiellopsis* as dothideaceous because of the thick stromatic wall. Waterman's figures (**1945**, figs. 2 A & B) definitely show that the asci are formed in separate cavities, separated from each other by interthecial strips of tissue. It differs from *Delphinella*, however, in the more definitely differentiated locule and the globose ascostromata. This is merely an example of the difficulty in placing intermediate types of ascostromatic development.

ATICHIACEAE:

The Atichiaceae is a debatable family which has been variously placed in the general scheme of the Ascomycetes, often as an "Anhang." They are usually found in the sterile or conidial condition. They occur on pine needles and leaves, mostly in the tropics. Several genera have been described, all of which are considered as synonyms of *Atichia* by Müller and von Arx (**1962**).

The fruiting bodies are made up of chains of sprout cells or "propagulae" of the form genus *Actinomma*. When moist the fruit bodies are gelatinous and slimy, but become cartilaginous when dry. They are tuberculate or lobed or with stellate arms. The asci are saccate with a thickened apical wall, and are in separate locules in *A. millardetii*, but occur in parallel groups in differentiated locules in *A. glomerulosa*. The family should probably be placed in the Myriangiales.

DOTHIDEALES

The order Dothideales is a confusing one. Fuckel (**1870**) and Lindau in Engler and Prantl (**1897**) included in it all of the stromatic forms. After von Höhnel's description of the Pseudosphaeriaceae, Theissen and Sydow (**1915**) in their monograph of the order limited it to those species with more than one locule. Those species with a single perithecium-like locule were placed in an order Pseudosphaeriales as an Anhang. These authors recognized four families, Polystomellaceae, Dothideaceae, Montagnellaceae and Phyllachoraceae. The Polystomellaceae were later placed in the Microthyriales because of their radiate shield. The Phyllachoraceae have been shown to have ascohymenial development and are transferred to the Sphaeriales. The Montagnellaceae were forms with a single locule and now are mostly found in the Pseudosphaeriaceae. Miller, Luttrell and others have pointed out that there are a number of genera which may have either one or many locules within a stroma. Luttrell (**1955**) accepted Miller's distinction which would place those species with interthecial parenchyma (variously termed paraphyses, pseudoparaphyses and paraphysoids in the older literature) in the Dothideales and those with true paraphysoids (in the sense of this work) in a new order Pleosporales. Both Theissen and Sydow (**1915**) and Clements and Shear (**1931**) included genera with and without paraphyses in their Dothideales. The term paraphysis at that time, however, did not distinguish between the different types of development of interthecial tissues. It must be admitted that the details of development of the interthecial tissues is known in only a few species of this group, and there seems to be a great deal of variation in this respect. Müller and von Arx did not recognize the Dothideales as an order and did not use the type of interthecial development as an ordinal character. They have distributed the stromatic genera of this order among a number of families, including the Botryosphaeriaceae, Parmulariaceae, Mycosphaerellaceae, Pseudosphaeriaceae, Venturiaceae and the Polystigmataceae.

There are some genera which may occur in either the unilocular or the plurilocular condition. There are, however, families and genera which are predominantly unilocular. Those with immersed ascostromata are placed in the Mycosphaerellaceae. Certain genera with an interthecial parenchyma and superficial unilocular ascostromata and an ecologic habit similar to the Dimeraceae and Erysiphales are placed in the families Englerulaceae and Perisporiopsidaceae. The Capnodiaceae are an artificial group based upon the character of the superficial mycelium.

A family key follows:

I. Superficial mycelium toruloid, often fused into rope-like strands ... CAPNODIACEAE
II. Superficial mycelium absent or not as above
  A. Ascostromata at least in part, plurilocular.................. DOTHIDEACEAE
  B. Ascostromata for the most part unilocular
    1. Ascostromata immersed in the substrate or erumpent.................... .... MYCOSPHAERELLACEAE
    2. Ascostromata superficial
      a. Ascostromata opening by a gelatinization of the upper tissues .... ENGLERULACEAE
      b. Ascostromata not gelatinous................. PERISPORIOPSIDACEAE

DOTHIDEACEAE:

There exists a number of genera which have a common stromatic structure. These genera are recognizable by the coarsely parenchymatic structure of the stroma. The cells of the stroma are isodiametric and often arranged in parallel rows. The outer layers of the stroma wall are composed of thick-walled pigmented cells which become thinner-walled and less pigmented toward the interior. These genera represent the original concept of the Dothideales. The development of the ascocarp is known in only a few species. The probabilities are that they do not all coincide with Luttrell's concept of the Dothideales as having a centrum of interthecial parenchyma. Some species show such interthecial parenchyma, but many others have the parenchyma compressed into filaments. They vary widely in other respects also, and have been placed in different families and orders on the basis of other characters. The stromata may be globose, pulvinate, effuse or on a basal stroma or hypostroma. They may be immersed erumpent or superficial. They may be unilocular or plurilocular. The locules may be entirely immersed in the stroma, in protruding outgrowths of a basal stroma, separate or confluent. Descriptions and figures in the literature show both interthecial parenchyma and filiform elements, which may be paraphysoids or paraphyses, in the centrum. As a result there is a confusing synonymy within the group and many genera have been transferred back and forth by different investigators. For convenience they are kept together here, until we can have much more information as to their ascostromatic development.

The following are adaptations of the keys to genera by Theissen and Sydow for the Dothideaceae and Montagnellaceae. They are entirely artificial but will serve as an index for a discussion of their distribution in other arrangements.

Dothideaceae

I. Ascostroma superficial with central hypostroma........................ Coccoideae
  A. Ascostroma with sterile centrum; loculi around it
    1. Spores 1-celled (4-celled?) hyaline, stroma smooth, paraphyses present...... .... YOSHINAGELLA
    2. Spores 2-celled, brown, stroma with peripheral tomentose hyphae, paraphyses present
      a. Loculi arranged peripherally......................... TRICHODOTHIS
      b. Loculi fused in a peripheral ring......................... PERISCHIZON
  B. Ascostroma fertile throughout
    1. Spores 1-celled
      a. Spores needle-like, hyaline; paraphyses present...... SCHWEINITZIELLA
      b. Spores elliptic hyaline; paraphyses absent.............. COCCOSTROMA
      c. Spores ellipsoid, brown; paraphyses present......... AUERSWALDIELLA
    2. Spores 2-celled
      a. Apiosporous, upper cell brown, ellipsoid, lower cell small, hyaline; paraphyses present.. COCCODISCUS
      b. Spores equally 2-celled
        i. Spores hyaline
          (a). Paraphyses present........................... MICROCYCLUS
          (b). Paraphyses absent .... COCCOIDELLA
        ii. Spores brown
          (a). Paraphyses present .... COCCODOTHIS
          (b). Paraphyses absent......................... COCCODOTHELLA
    3. Spores 3-celled, hyaline; paraphyses present................ COCCODIELLA
    4. Spores 4-celled, hyaline; paraphyses present............. ELMEROCOCCUM
II. Ascostroma superficial with many hypostromata....................... Leveillelleae
  A. Spores 2-celled, hyaline; paraphyses absent................. MICROCYCLELLA
  B. Spores 2-celled, brown
    1. Paraphyses present
      a. Ascostromata smooth; free mycelium present............. LEVEILLELLA
      b. Ascostromata smooth; no free mycelium............. ENGLERODOTHIS
    2. Paraphyses absent
      a. Ascostromata smooth; no free mycelium .... LEVEILLINA .... POLYRHIZON
      b. Ascostromata with conidial hairs; no free mycelium .... DISCODOTHIS
  C. Spores filiform.. TRICHOCHORA
III. Ascostromata immersed, erumpent .... Dothideae
  A. Spores 1-celled
    1. Spores hyaline
      a. Spores allantoid (valsoid)......................... DOTHIDEOVALSA
      b. Spores ellipsoid
        i. Paraphyses present
          (a). Asci 3-4-spored .... ZIMMERMANNIELLA
          (b). Asci 8-spored.............................. BAGNISIOPSIS
        ii. Paraphyses absent
          (a). Ascostroma regular, pulvinate.............. BOTRYOSPHAERIA .... (AMERODOTHIS)
          (b). Locules in erect cylindric outgrowth............... CATABOTRYS
    2. Spores brown
      a. Paraphyses absent................................ AUERSWALDIA
      b. Paraphyses present .... DOTHIDINA

B. Spores 2-celled
  1. Spores hyaline
    a. Paraphyses present .... ULEODOTHIS
    b. Paraphyses absent .... DOTHIDELLA
  2. Spores brown
    a. Paraphyses absent .... SYSTREMMA
    b. Paraphyses present .... ACHORELLA
C. Spores 3-celled, hyaline .... METAMERIS
D. Spores 4-celled
  1. Spores hyaline; paraphyses absent .... PHRAGMODOTHELLA
  2. Spores brown, paraphyses present .... PHRAGMODOTHIS
E. Spores muriform, brown; paraphyses absent .... DICTYODOTHIS

Montagnellaceae

I. Loculi single, immersed or erumpent or on a basal stroma .... Montagnelleae
  A. Spores 1-celled, hyaline
    1. Without a definite stroma; paraphyses present .... HAPLOTHECIUM
    2. Loculi sunken in a stroma; paraphyses absent .... MONTAGNELLINA
  B. Spores 2-celled
    1. Spores hyaline; paraphyses absent
      a. Without a definite stroma
        i. Loculi single, scattered, immersed, with a thick upper wall ....
        .... HAPLODOTHIS
        ii. Loculi confluent, erumpent, not thickened above ....
        .... PSEUDOSPHAERELLA
      b. With a stroma
        i. Loculi irregularly scattered .... DIPLOCHORELLA
        ii. Loculi arranged in a ring .... CYCLODOTHIS
        iii. Loculi arranged in lines .... SCIRRHIACHORA
    2. Spores brown
      a. Without a definite stroma; paraphyses present .... PYRENOBOTRYS
      b. With a stroma; paraphyses absent
        i. Ascostromata irregularly distributed .... CROTONE
        ii. Ascostromata in branching lines .... TERATOSPHAERIA
  C. Spores 4-celled, brown, with stroma
    1. Paraphyses present .... SYNCARPELLA
    2. Paraphyses absent .... MONTAGNELLA
  D. Spores muriform; with stroma
    1. Spores hyaline; paraphyses absent .... HYALOCURREYA
    2. Spores brown
      a. Paraphyses present .... CURREYA
      b. Paraphyses absent .... EPIBOTRYS
  E. Spores filiform; paraphyses absent .... OPHIOCARPELLA
II. Loculi superficial on an immersed basal stroma, or in special stromatic columns....
.... Rosenscheldieae
  A. Loculi superficial in a subcuticular basal stroma; paraphyses absent
    1. Crowded on a common stroma
      a. Spores 2-celled, hyaline .... ROSENSCHELDIELLA
      b. Spores 2-celled, brown .... LICOPOLIA
    2. Stromata simple, perithecium-like with a short immersed foot; spores 2-celled, hyaline .... MONOPUS

B. Loculi single on an erumpent stromatic column
  1. Spores 4-celled, brown; paraphyses present .... ROSENSCHELDIA
C. Stroma cylindric, erumpent with a grape-like cluster of loculi
  1. Spores 2-celled, brown; paraphyses present................ STALAGMITES

Von Arx and Müller (**1954**) and Müller and von Arx (**1962**) have distributed the genera included here under the Dothideales because of their characteristic stromatic tissues in a number of families, on the basis of other characters. Some of these arrangements are indicated here:

Botryosphaeria, Auerswaldia, Auerswaldiella, and *Montagnellina* are placed in the Botryosphaeriaceae because of their one-celled ascospores. *Auerswaldia*, erumpent on bark and *Auerswaldiella* on leaves have a foot or hypostroma within the host. *Montagnellina* has single locules imbedded in an effuse sub-epidermal ascostroma.

Botryosphaeria: The genus *Botryosphaeria* has erumpent ascostromata which are greatly variable in form. They may be plurilocular, in which case the locules are entirely immersed in a pulvinate stroma, or the locules may occur singly in globose or stalked outgrowths of a basal stroma. Locules may also occur singly in scattered or crowded ellipsoid ascostromata. In this latter case the species have formerly been placed in the genus *Physalospora*. Von Arx and Müller (**1954**) include such species in the genus *Botryosphaeria* and limit the genus *Physalospora* to ascohymenial species in their Polystigmataceae.

The conidial states of *Botryosphaeria* are likewise variable, but always occur in pycnidial locules, either in stromata identical to the ascostromata or even in the same stroma as the ascus locules. The conidia are at first ellipsoid, hyaline and one-celled, but may become brown and in some instances one-septate. If all types of conidia occur in plurilocular stromata they fall in the form genus *Botryodiplodia*, if only one-celled, hyaline conidia, in the form genus *Dothiorella*. If unilocular pycnidia have only the hyaline spores they fall in the form genus *Macrophoma*, if the spores are one-celled and brown, in the form genus *Sphaeropsis*, and if the conidia are two-celled and brown in the form genus *Diplodia*. Bacillar-microconida, also in pycnidia, have been reported by Shear and Stevens (**1924**).

Grossenbacher and Duggar (**1911**) described a shoot blight of *Ribes*, caused by *Botryosphaeria ribis*. They found two strains of this species. One was chromogenic, forming a purple-pink color on starch paste and was strongly parasitic. The other was non-chromogenic and grew only as a saprophyte. Stevens (**1934**) and Shear and Stevens (**1924**) studied the black rot of apple, caused by *Sphaeropsis malorum*, the conidial state of *Physalospora malorum* (*P. cydoniae*). This species was found to occur mixed with *B. ribis* on a number of woody hosts including *Ribes*.

Von Arx and Müller (**1954**) recognized two species of *Botryosphaeria* in this group. *Botryosphaeria quercuum*, which is common as a twig blight of oak, and in which they included *Physalospora malorum*, has large ascospores, 23-38 × 7-13 μ and a *Botryodiplodia* or *Sphaeropsis* conidial state; *B. dothidea* (*B. ribis*) has smaller ascospores, 13-26 × 5-10 μ and a *Dothiorella* conidial state. They also included *Physalospora mutila*, with a conidial state, *Diplodia mutila*, which occurs on apple in England (Stevens, **1936**), in *B. quercuum*. The interthecial tissue of both these species has been described by Shear and Stevens (**1924**) as being formed by the compression of the stroma parenchyma into filiform strips.

Botryosphaeria piceae (Funk, **1965**) is reported as parasitic on the branches of spruce and *B. tsugae* (Funk, **1964**) as a die back of western hemlock.

Delphinella (*Rehmiellopsis*), as described in the Dothioraceae has asci separated by interthecial tissue in a simple locule, within a globose ascostroma. It differs from *Mycosphaerella* chiefly in the polysporous asci, and is intermediate between the Dothioraceae and the Mycosphaerellaceae.

Perischizon, Coccodothis, Englerodothis, Rhagadolobium (*Discodothis*) and *Hysterostomella* are placed in the Parmulariaceae, because of the presence of a hypostroma. *Englerodothis, Rhagadolobium* and *Hysterostomella* have a more or less effuse ascostroma with one or several hypostromata anchoring it to the host. They have irregular disc-shaped locules with a radiate upper covering. They are similar to the Microthyriales and have a discomycete-like structure. *Perischizon* has locules arranged in a circle about a central sterile plug. *Coccodothis* has apiosporous ascospores and grouped locules in a pulvinate erumpent ascostromata with a central hypostroma.

Licopolia and *Polyrhizon* are placed in the Pleosporaceae, a family consisting mostly of genera with unilocular ascostromata and supposedly having an interthecial tissue of paraphysoids. *Licopolia* has unilocular, globose ascostromata upon a parenchymatic basal stroma. *Polyrhizon* with plurilocular ascostroma on an intraepidermal hypostroma, is figured by Müller and von Arx (**1962**, fig. 120) as having interthecial parenchyma.

The Mycosphaerellaceae of Müller and von Arx (**1962**) also contains mostly unilocular genera. They are distinguished from the Pleosporaceae by the presence of interthecial parenchyma in the locules. *Microcyclus, Uleodothis* and *Rosenscheldiella* have superficial ascostromata with basal hypostromata. *Rosenscheldiella* has unilocular ascostromata, more or less fused together and seated on an internal hypostroma. *Uleodothis* is figured (**1962**, fig. 149) as having interthecial paraphysoids. *Dothidea* and

Achorodothis have apiosporous ascospores and *Scirrhia* has subepidermal, effuse stromata.

The Venturiaceae of Müller and von Arx is based upon the ascospores which are yellowish to olive to brown and mostly apiosporous. *Trichodothis, Coccoidea, Coccodiella. Rhizogene, Atopospora* and *Crotone* are all plurilocular genera with superficial ascostromata with a basal foot or hypostroma in the host. *Coccoidella, Trichodothis* and *Rhizogene* have ascospores with a septum in the middle or above (not apiosporous). *Rhizogene* has locules in the periphery of a sclerotioid stroma with hairs or setae externally.

Dibotryon: The genus *Dibotryon* was erected by Theissen and Sydow (**1915**) for the fungus causing the black knot of species of *Prunus*. *Dibotryon morbosum* is systemic and perennial. According to Dandero (**1907**) it takes three years to form mature asci. In the first year mycelium penetrates throughout the host causing an unsightly knot-like hypertrophy. A light brown parenchymatic stroma is formed which pushes through breaks in the bark. One-celled, brown *Cladosporium*-like conidia are cut off from short stout conidiophores on the surface of this stroma. In the second year the stroma turns black on the surface and produces numerous papillate outgrowths, which do not, however, contain any asci. In the third year mature asci are formed in locules within these papillate outgrowths. The hyaline, unequally two-celled ascospores are distributed the third spring. Müller and von Arx (**1962**) give *Dibotryon* as a synonym of *Apiosporina* in their Venturiaceae, but the two genera are entirely different except for the apiosporous ascospores.

Dothidea: Luttrell (**1951**) gave a detailed account of the development of *Dothidea collecta*. The stromata are formed on a number of woody hosts. The asci contain four, unequally two-celled, brown ascospores. Young stromata arise at the end of January as a mass of multinucleate parenchyma cells beneath the periderm. In culture conidial locules were formed in the surface of the young stroma. In the formation of such locules a circular mass of deeply staining cells undergoes a division of their protoplasts. Each of these "mother cells" produce a sterigma upon which a bacilliform conidium is formed. Shortly thereafter (in February) ascogonia appear in the stroma beneath the pycnidia. Each ascogonium arises from a stroma cell which takes a deeper stain. A hypha (trichogyne) is pushed upward from the ascogonial cell, sometimes to the surface of the stroma, where it forms a bulbous tip. No fusion of microconidia with such trichogynes was observed. Shortly after the origin of the ascogonia, the ascogenous locules arise. An ellipsoidal mass of deeply staining cells, with very thin walls, is differentiated within the stroma to form each locule. There were more ascogonial cells than locules, but each locule contained one or more ascogonia. During the early summer these ascogonia disappeared. In June, a small group of deeply staining cells appeared in the base of the locules. These cells arose from the segmentation of the multinucleate protoplasts. These segmented cells became the binucleate ascogenous cells from which ascogenous hyphae arose, to form a basal cushion from which the asci arose directly without the intervention of croziers. No direct connection between the ascogonia and the ascogenous cells could be detected. As the asci developed, the protoplasts of certain cells of the locule parenchyma, above the ascogenous hyphae, fragmented. These cells were eventually pushed back and disintegrated by the developing asci. The asci were bitunicate. This is the typical development within the Dothideales.

Hess and Müller (**1951**), in their study of the development of the ascostroma of *Dothidella* (*Dothidea*) *insculpa*, reported that there were male "antheridial cells" within the locule parenchyma which could fuse with the trichogynal outgrowths from the ascogonia. They saw the fragmentation of the protoplasts of parenchyma cells, but termed the fragments macroconidia, which they believed could also fuse with a trichogyne. They saw a similar formation of microconidia in irregular cavities within the stroma, for which they erected the genus *Asteromellopsis.*

Loeffler (**1957**) in his study of the genus *Dothidea*, reported *Asteromellopsis* conidial states for a number of species of the genus. When conidiophores form a hymenium from the surface cells of the stroma, they are placed in the form genus *Systremmopsis.*

In *Dothidella ulmi,* (*Platychora* Petrak) Killian (**1920**) found that a subepidermal conidial stroma was first formed on the upper surface of the leaves. A parallel layer of pseudoparenchyma, on the surface of the stroma, cut off conidia which were obpyriform to bottle-shaped, bicellular, and brown. Ascostromata arose as parenchymatic stromata beneath the conidial stromata. Within such stromata he observed certain cells which enlarged, became deeply staining and divided several times. Two of these cells fused. He termed them male (antheridium) and female (ascogonium). Nuclei passed from the antheridium into the ascogonium where they paired and passed out into the ascogenous hyphae as outgrowths from the ascogonium. The asci matured the following spring. He showed interthecial tissue of somewhat elongated parenchyma cells of the locule.

Dothidella ulei is the cause of a leaf blight of the rubber tree (*Hevea brasiliensis*) in South America. It appears first on young leaves as a conidial state (*Passalora heveae*), followed by a pycnidial state (*Aposphaeria ulei*) and finally by the ascigerous state. It occurs as a leaf spot, but also on petioles, stem cankers, and fruits.

Systremma natans (*Dothidea sambuci* accd. Müller & von Arx) was studied by Luc (**1952**). The ascostromata are disc-like with an outer layer of dark parenchyma cells and similar inner cells with hyaline walls. The locules arise by the division of these stroma cells to form an ellipsoid area of small thin-walled cells. A few cells in the locule remain large and become the ascogonia, which, in turn, give rise to the ascogenous hyphae. The asci arise from these hyphae and push upward between and dissolve away the locule parenchyma. No sterile interthecial tissue remains. The spores are somewhat unequally two-celled and brown.

The genera *Scirrhia, Systremma* and *Cymadothea* are all similar in that they form effuse subepidermal ascostromata which are often made up of confluent or fused unilocular ascostromata.

Scirrhia (*Systremma*) *acicola* and *S. pini* are the causes of a brown spot and needle blight of yellow pines. *Scirrhia acicola* was studied by Wolf and Barbour (**1941**). It forms conidial stromata in the mesophyll which have a surface palisade of conidiophores which bear cylindric to fusoid, one- to three-septate, brown conidia. During winter or spring, dark colored spherical to elongate ascostromata arise, in which one to fourteen locules are formed. Some of these locules become spermagonia and produce bacillar spermatia. The ascus locules contain a loose parenchyma, within which an ascogonium with an elongate trichogyne extends above the surface. The asci are fasciculate and there are no interthecial strips. The conidial state is known as *Lecanosticta acicola. Scirrhia pini* has a somewhat similar conidial state, but the conidia are hyaline and they are placed in the form genus *Dothistroma*. The ascostromata consist of a linear series of fused locules. The interthecial tissues consist of chains of globose cells ("paraphyses") (Funk & Parker, **1966**). Microconidial spermagonia with rod-shaped conidia were also seen in this species.

Cymadothea trifolii was described under this genus name by Wolf (**1935**). It is usually found in the conidial state, which causes a sooty blotch of various species of *Trifolium*. It has been placed in various genera (*Dothidea, Plowrightia. Dothidella, Mycosphaerella*) (Petrak, **1941** in *Mycosphaerella*) by supposition. Von Arx (**1949**) made *Cymadothea* a subgenus of *Mycosphaerella*, because the fruiting pustules consist of fused unilocular ascostromata. The conidial state occurs throughout the summer as a sooty mycelial growth which forms fascicles of palisade-like conidiophores through the epidermis. Two-celled, brown conidia are cut off from the apex of the conidiophore. A second conidium is then formed on a side branch of the conidiophore which pushes aside the first conidium. The formation of a succession of such conidia, gives rise to a zig-zag conidiophore with a conidial scar at each angle of the zig-zag. This conidial state has been known as *Polythrincium trifolii*. During autumn, spermagonia arise as spherical to pyriform cavities in a subepidermal stroma. The spermatia are bacilliform, hyaline and are exuded in large numbers. At about the same time, ascus locules arise in these same stromata as spherical areas of deeply staining, hyaline thin-walled cells. Some cells in the base of these locules remain large and grow upward as trichogynes which extend to the surface, where they may fuse with spermatia. The formation of ascogenous hyphae, origin of the asci and fusion of the nuclei in the ascus, takes about six weeks. The ascospores, which are somewhat unequally two-celled, and hyaline to pale yellow, mature the following spring. Wolf placed *Cymadothea* in the Dothideales near *Systremma*, because of the dark brown ascostroma. This is an example of one of the intermediate forms between the Dothideaceae and the Mycosphaerellaceae, as here outlined.

Obrist (**1959**) found that *Scirrhia* (ascospores two-celled), *Scirrhophragma* (ascospores two- or four-celled) and *Rhopographus* (ascospores three- to seven-septate) all form effuse, linear, subepidermal ascostromata on ferns and grass culms. They form an *Acremonium* type of one-celled, ellipsoid conidia on lateral conidiphores upon the mycelium. They may also form spherical conidia, cut off from an hymenium on the surface of young stromata. These fall in the form genus *Hadrotrichum*.

In *Dangeardiella*, he found that the ascostroma, if plurilocular, was formed by the fusion of unilocular stromata. They have a longitudinal slit-like opening, as in the Lophiostomataceae. The spores are many-septate.

The genera *Bagnisiopsis, Coccostroma, Zimmermanniella, Dothidina* and *Sucinaria* are all very similar leaf parasites with superficial uni- or pluri-locular stromata, with a basal hypostroma. The spores are hyaline becoming brown with age. These genera have been placed in the Sphaeriales by Petrak (**1928**), the Polystigmataceae by von Arx and Müller (**1954**) and in the Dothideales by Miller and Burton (**1943**). The stromata are mostly pseudoparenchymatic as in the Dothideales, but are pale colored within in some species. Authors differ in the interpretation of the locule wall, as to whether it is merely a differentiated layer of the ascostroma or arises as a true wall as in the ascohymeniales. Its true origin has not been determined. The asci are thin-walled and are separated by "paraphysoids." In *Succinaria* and in some species of *Bagnisiopsis* the internal stroma is light in color and these have been placed in the Hypocreales. Miller and Burton (**1943**) described conidial states in the form genus *Hemidothis* in several species of *Bagnisiopsis*. The conidia are filiform, curved, hyaline and borne in cavities. *Zimmermaniella* has asci with three-four ascospores.

Moreau has described the development of *Montagnella opuntiarum*. The stroma forms a mass of pale brown parenchyma, erumpent through the periderm. The central locules are differentiated by the division of the ascostroma cells to form a small celled hyaline parenchyma. Ascogenous hyphae arise in the base of the locule and give rise to a fascicle of asci which dissolve their way through the locule parenchyma. The ascospores are four-six-eight in an ascus and are long fusoid hyaline, two-celled at first, becoming yellowish and four-celled at maturity.

From the above discussion of genera it can be seen that there is a good deal of difference of opinion as to their placement. The following key, although admittedly artificial is given as a convenient index for the confused student who wishes to locate genera in the Dothideaceae as here circumscribed. It includes the genera with coarse parenchymatic tissue. They are for the most part multilocular, although some genera may also form unilocular stromata.

I. Spores 1-celled
  A. Stromata superficial with a basal hypostroma within the host
    1. Erumpent through the bark on woody stems .... AUERSWALDIA
    2. Erumpent through the epidermis, on leaves .... AUERSWALDIELLA
  B. Stromata immersed or erumpent
    1. Stromata subepidermal, on leaves, locules scattered ...... MONTAGNELLINA
    2. Stromata various, erumpent on stems, locules immersed in a basal stroma, within outgrowths of such a stroma or as separate unilocular stromata .... BOTRYOSPHAERIA
II. Spores 2-celled
  A. Ascostroma erumpent with a central hypostroma within the host
    1. Stromata with marginal locules
      a. Stromata with a hyphal tomentum
        i. Ascostromata flat, dimidiate....................... TRICHODOTHIS
        ii. Ascostromata disc-shaped or pulvinate, raised on a foot or mycelium
          (a). Locules immersed in margin .... RHIZOGENE
          (b). Locules superficial, beneath the ascostromata...... LASIOBOTRYS
      b. Stromata without tomentum, locules small .... XENOMERIS
    2. Stromata with locules not marginal
      a. Ascostromata flat, dimidiate, locules around a central, sterile column
        i. Locules in a ring around a sterile centrum............. PERISCHIZON
        ii. Locules radiating from a central sterile column ... SYMPHAEOPHYMA
      b. Ascostromata effuse, disc-like or pulvinate, locules scattered throughout the stroma
        Ascospores equally 2-celled
          (a). Asci few, fasciculate; spores yellow-brown......... COCCOIDELLA
          (b). Asci numerous, parallel; spores hyaline to yellow-brown......... .... ULEODOTHIS
        ii. Ascospores unequally 2-celled (apiosporous)
          (a). Ascostromata effuse on an effuse hypostroma .... POLYRHIZON
          (b). Ascostromata strongly erumpent, globose to pulvinate
            (1). Ascostromata hemisphaeric .... COCCODOTHIS
            (2). Ascostroma disc-like......................... COCCOIDEA
            (3). Ascostroma pulvinate, on hypertrophy of the host .... CROTONE
        iii. Ascospores equally 2-celled or sometimes apiosporous
          (a). Ascostromata various, erumpent, locules immersed. MICROCYCLUS

(b). Ascostroma immersed, locules in distinct, superficial, confluent outgrowths . . . . ROSENSCHELDIELLA
(c). Ascostromata various, in distinct superficial outgrowths, often setose . . . . GIBBERA

B. Ascostromata without a basal hypostroma within the host
  1. Ascospores unequally 2-septate (apiosporous)
    a. Subcuticular, ascostromata effuse
      i. Locules scattered in stroma . . . . OMPHALOSPORA
      ii. Locules crowded in stroma . . . . ATOPOSPORA
    b. Ascostroma intraepidermal, effuse . . . . PLATYCHORA
    c. Ascostromata subepidermal, erumpent, on hypertrophy of the host
      i. Stroma thick, within the bark, erumpent, locules in papillate swellings . . . . DIBOTRYON
      ii. Ascostromata on a hypertrophy of leaves, locules immersed . . CROTONE
  2. Ascospores equally 2-celled
    a. Ascostroma subcuticular, crustose . . . . EURYACHORA
    b. Ascostromata subepidermal, linear
      i. On grasses and ferns, effuse . . . . SCIRRHIA
      ii. On leaves of higher plants, ascostromata of fused locules . SYSTREMMA
      . . . . CYMADOTHEA
    c. Ascostromata erumpent, locules in superficial outgrowths . . . . ROSENSCHELDIELLA

III. Spores more than 2-celled (phragmosporous)
A. Superficial with a central hypostroma in the host
  1. Ascospores 3-celled, hyaline . . . . COCCODIELLA
  2. Ascospores 4-celled, hyaline . . . . ELMEROCOCCUM
B. Ascostromata immersed with a superficial clypeus
  1. Acostromata subcuticular
    a. Ascospores 3-celled, brown . . . . DERMATODOTHIS
    b. Ascospores 4-celled, hyaline . . . . CAMAROTELLA
  2. Ascostromata subepidermal
    a. Ascostromata circular, spores 4-celled, hyaline . . . . PHRAGMOCAUMA
    b. Ascostromata linear, effuse
      i. Ascospores 3-celled, hyaline . . . . SCIRRHOPHRAGMA
      ii. Ascospores 4- to many-celled, hyaline . . . . EXARMIDIUM
      iii. Ascospores 4- to many-celled, brown. . . . . RHOPOGRAPHUS
  3. Ascostromata immersed in the mesophyll
    a. Ascospores 4-celled, hyaline
      i. Paraphysoids present. . . . . TELIMENA
      ii. Paraphysoids lacking . . . . PHRAGMOCARPELLA
    b. Ascospores 4-celled, brown
      i. Paraphysoids present . . . . HOMOSTEGIA
      ii. Paraphysoids lacking. . . . . CLYPEOSTROMA
C. Ascostroma immersed, erumpent (no clypeus); spores 4-celled, brown
  1. Locules in slightly swollen papillae, erumpent
    a. Paraphysoids present. . . . . SYNCARPELLA
    b. Paraphysoids lacking . . . . MONTAGNELLA
  2. Locules in superficial outgrowths of a basal stroma. . . . . ROSENSCHELDIA

IV. Spores muriform (dictyosporous)
A. Ascostromata immersed beneath a clypeus, spores brown
  1. Paraphyses present . . . . DICTYOCHORELLA
  2. Paraphyses absent. . . . . DICTYOCHORA

B. Ascostromata erumpent (no clypeus)
  1. Loculi crowded in or on a sunken stroma
    a. Ascospores hyaline .... HYALOCURREYA
    b. Ascospores brown, paraphyses present .... CURREYA
    c. Ascospores brown, paraphyses absent .... EPIBOTRYS

V. Spores long, filiform (scolecosporous)
  A. Ascostromata superficial with a central foot............... SCHWEINITZIELLA
  B. Ascostromata superficial, with several hypostromata .... TRICHOCHORA
  C. Ascostromata immersed beneath a clypeus
    1. Ascostromata subepidermal.......................... SCOLECODOTHIS
    2. Ascostromata immersed in the mesophyll............... OPHIODOTHELLA
  D. Ascostromata immersed, without a clypeus .... OPHIOCARPELLA

MYCOSPHAERELLACEAE:

The family Mycosphaerellaceae was originally used for forms with simple immersed "perithecia" which had no paraphyses, in contradistinction to the Pleosporaceae, which did have paraphyses. Müller and von Arx recognized the same two families under their order Pseudosphaeriales, the Mycosphaerellaceae without, and the Pleosporaceae with, "paraphysoids." It is not clear what they mean by paraphysoids. Under the Mycosphaerellaceae they say "Das zarte, aus hyalinen eckigen Zellen bestehende Binnengeflecht wird durch die sich entwickelnden Asci resorbiert oder teilweise nach oben und an den Rand gedrückt. Daher fehlen in den reifen Loculi die Paraphysoiden. In grösseren Fruchtkörpern stehen die Asci mehr oder weniger parallel nebeneinander, und das senkrecht zellige oder hyphige Binnengeflecht bleibt in Form von Paraphysoiden teilweise erhalten." From their figures they apparently include ascostromata which at maturity have either unaltered locule parenchyma (or none) (*Mycosphaerella, Scirrhia*) or filiform strips of parenchyma formed by compression from the asci (*Uleodothis, Rosenscheldiella*). This agrees with our present delimitation of the Mycosphaerellaceae, for there are many intermediate situations between the disintegration of the unaltered locule parenchyma and the filiform strips formed by compression from the asci. However, Müller and von Arx include both immersed and superficial, and unilocular or multilocular ascostromata in their family, whereas the following group of genera are limited to the immersed unilocular forms.

There are certain genera which occur in both the unilocular and the multilocular condition. In *Botryosphaeria* the unilocular form was formerly placed in the genus *Physalospora*. In *Cymadothea, Systremma,* etc., which von Arx (**1949**) makes a subgenus of *Mycosphaerella,* the fruit bodies may occur as separate unilocular stromata or as fused linear groups forming a multilocular ascostroma. Such genera are here considered under the Dothideaceae.

I. Spores 1-celled, ascostromata immersed, without an ostiole
  A. Asci sessile, cylindric, lying parallel to one another
    1. With a thickened ring of wall tissue about the ostiole ..... DISCOSPHAERINA
    2. Without such a thickened ring .... GUIGNARDIA
  B. Asci stipitate, clavate, lying at different levels................ VESTERGRENIA
II. Spores 2-celled
  A. Spores hyaline
    1. With dark setae in the ostiole .... KEISSLERIELLA
    2. Without setae in the ostioles
      a. Asci clavate, fasciculate, spores not constricted at the septum, no filiform strips .... MYCOSPHAERELLA
      b. Asci cylindric, parallel, ascospores constricted at the septum, with filiform interthecial strips................................ DIDYMELLA
    1. Ascostromata hemispherical, with a flattened base, superficial or erumpent
      a. Ascospores less than 40 μ long, only 2-celled .... MICROTHELIA
      b. Ascospores over 40 μ long, with incomplete septa in the end cells........ .... CARYOSPORA
    2. Ascostromata spherical, immersed
      a. Ascostromata immersed in a secondary hyphal stroma, ostiole elongate .. .. SYDOWINA
      b. Ascostromata not in such a secondary stroma, ostiole papillate .... DIDYMOSPHAERIA
III. Spores 2- to many-celled
  A. Asci saccate to broad clavate, few in number (1-6), apical wall thickened
    1. Ascostromata small, 80-120 μ in diameter, spores not over 30 μ long .... LEPTOSPHAERULINA
    2. Ascostromata larger, 120-700 μ in diameter, spores more than 30 μ long..... .... WETTSTEININA
  B. Asci clavate to cylindric, more numerous
    1. Ascospores 3-celled, with the central cell larger........ LEPTOGUIGNARDIA
    2. Ascospores many-celled, nearly equal in size................ SPHAERULINA

Guignardia: The genus *Guignardia* occurs, mostly, as a parasite on leaves or stems. It forms small spherical subepidermal ascostromata, which have a coarse parenchymatic wall and no ostiole. The interthecial tissue remains unaltered or is dissolved away by the asci which contain ellipsoid, one-celled, hyaline spores.

The genus *Discosphaerina* has been used by Höhnel and Petrak for species of *Guignardia* which have a ring-like thickening on the upper wall. This genus is doubtful and is made a synonym of *Guignardia* by von Arx and Müller (**1954**).

Both microconidia and macroconidia are formed by species of *Guignardia*. Microconidia are formed in pycnidia of the form genus *Asteromella*. The macroconidia have been placed in numerous form genera by various writers (Müller, **1957, 1959**; Conners, **1959, 1959a**; Reusser, **1964**). Müller has reduced the many similar form genera to the following: *Phyllostictina*, with one-celled, hyaline, ellipsoid spores in membranous pycnidia; *Selenophoma*, with one-celled lunate spores in pycnidia; *Dothiorella*, with one-celled ellipsoid spores in stromatic pycnidia; *Colletotrichella*, with one-celled ellipsoid conidia in open acervuli; and *Kabatia*, with two-celled, lunate conidia in flat acervuli. All of these types of conidia may occur free upon the mycelium, in culture. Species with apparently identical ascostromata, may give rise to different conidial states, suggesting the presence of host species or physiologic strains.

Guignardia bidwellii is the cause of a black rot of grapes. It attacks the grape berries and reduces them to blackened, wrinkled mummies, upon which the pycnidial state, *Phyllostictina* (*Phoma*) *bidwellii* occurs in great numbers. The ascostromatic state may occur later on the berries or also upon the leaves.

Several doubtful species of *Guignardia* have been described as parasites of marine algae, forming "composite" plants which have been compared to lichens. *Guignardia alaskana* and *G. prasiolae* are found on *Prasiola* and *G. ulvae* on *Ulva.* These species are given as having an ostiole.

Vestergrenia: The species of this genus are leaf parasites. The ascostromata have no ostiole and are immersed in the host beneath a development of a subepidermal clypeus. The asci become long stipitate and then lie in a cluster at different heights within the locule. The spores are biseriate, hyaline and one-celled.

Mycosphaerella: The genus *Mycosphaerella* is a very large one, of some one thousand described species. Most of these are leaf parasites, causing the so-called leaf-spots. Many species, however, have been described upon the basis of host occurrence, and many should be considered as synonyms or physiologic races or large specific groups.

In the typical life history within the genus, infection occurs on young leaves in early spring. The host tissues are killed within a limited area causing necrotic, brown spots on the leaf. During the summer, the conidial state occurs on these spots. Towards fall spermogonia may be found and ascostroma initials appear. In most species the ascostromata mature over winter in a saprophytic environment of growth. The ascospores mature and are disseminated the following spring.

A large variety of conidial forms has been connected with species of *Mycosphaerella* (Klebahn, **1918**; von Arx, **1949**). The hyphomycetous types occur as small pustules of erumpent conidiophores on the leaf spots during the summer. In the form genus *Ovularia* the conidia are one-celled, spherical and hyaline; in *Ramularia* the conidia are fusoid to cylindric, two- or many-celled, and hyaline, and borne in chains; in *Cercosporella* the conidia are more elongate many-septate and hyaline; in

Cercospora the conidia are as in *Cercosporella* but they and the conidiophores are pigmented; in *Passalora* the conidia are as in *Cercospora*, but shorter and two-celled; in *Polythrincium* the conidia are ellipsoid, one- to two-celled and borne on zig-zag conidiophores; in *Cladosporium* and *Scolecotrichum* the conidia are one-celled, or septate, brown, borne singly or in chains on short conidiophores.

Conidia also occur in pycnidia. In the form genus *Septoria,* the conidia are long-cylindric to filiform and one- to many-celled, hyaline or colored (*Phaeoseptoria*). In *Phleospora* the conidia are similar but are found in pycnidia which are wide open at maturity.

Microconidia occur in late summer and fall, and in some cases at least bring about fertilization. The microconidia are small, rod-like and produced in large numbers in immersed, *Phyllosticta*-like pycnidia which have been given such form generic names as *Asteromella, Stictochorella,* and *Plectophoma.*

Von Arx (**1949**) in his study of the genus *Mycosphaerella* recognized three subgenera. *Eumycosphaerella* has fasciculate asci which are rather thin-walled, clavate to cylindric and often discharged as a united fascicle. *Didymellina* has more saccate asci, swollen at the base and with a thickened apical wall, and are discharged singly. *Cymadothea* includes species in which the unilocular ascostromata are often fused together to form a stroma.

Mycosphaerella tassiana, the type species of the subgenus *Didymellina*, occurs on the leaves and stems of many plants, both monocotyledons and dicotyledons, in the subarctic and subalpine regions. This species was studied in culture by Barr (**1958**). In culture at room temperature, only the conidial *Scolecotrichum* state was formed. If, after a week or two at room temperature, the cultures are removed to 5 C, ascostromata initials will be formed within four weeks. These arise by the enlargement and division of certain cells within the mycelium. In the base of these stromata certain cells enlarge, become deeply staining and undergo nuclear division. These cells put out hyphae which become the ascogenous hyphae. The asci arise directly from the apical cells of these hyphae or through the medium of croziers. After nine-ten weeks in the cold room at 5 C, fusion nuclei could be seen. The asci may remain in this condition for some seven weeks. If the cultures are removed from the cold room at various periods, this succession of events is hastened. In nature the asci remain in fusion nucleus stage until early spring, when meiosis and the formation of ascospores takes place. The asci dissolve their way through the locular parenchyma. No spermagonia were formed. This necessity of a cold period is apparently true for many parasitic species which develop their ascostromata over winter in a saprophytic environment, and disperse their ascospores the following spring.

Higgins (**1920, 1929, 1936**) studied the development of *Mycosphaerella bolleana, M. personata* and *M. tulipiferae*, of the subgenus *Eumycosphaerella.* The life history was similar in all of these leaf parasites. Each of the three formed a *Cercospora* conidial state during the summer on the leaf spots. In August spermagonia arose by the branching and coiling of vegetative hyphae to form a spherical, subepidermal mass of parenchyma. Central cells in this stroma took a deep stain, increased in size and underwent two mitoses to form a four-nucleate cell. The protoplast of such cells then divided to form four uninucleate protoplasts, or "spermatia mother cells." Each of these cells then produce a spermatium upon a sterigma.

Ascostromata arose in the same way as the spermagonia. A deep staining cell in the base of the ascostroma became an ascogenous cell and sent up a hyphal trichogyne which extended through the epidermis. Spermatia became attached to this trichogyne and supposedly passed a nucleus downward to the ascogenous cell, which became binucleate. The developing asci pushed their way through the hyaline locule parenchyma cells and disintegrated them entirely. These species differ from *M. tassiana* in the formation of spermagonia and the fertilization of the ascogenous cell by way of the trichogyne, and in the *Cercospora* conidial state.

Similar life cycles have been reported for a number of species of *Eumycosphaerella* by F.A. Wolf (**1935, 1935a, 1936, 1938, 1938a, 1940, 1943**) and by Wolf and Barbour (**1941**).

Mycosphaerella tassiana is given by Müller and von Arx (**1950**) as the origin of one stem of their Loculoascomycetes (see under Classification).

Mycosphaerella ascophylli is found on the marine alga *Ascophyllum* and *M. pelvetiae* on *Pelvetia.* Johnson and Sparrow (**1961**), in their description of the genus *Mycosphaerella* give the asci as being sixteen-spored, but these two species as having eight-spored asci.

Didymella: The genus *Didymella* is very similar to *Mycosphaerella* and differs only in degree (Corbaz, **1957**). The ascostromata are usually larger than in *Mycosphaerella*, and in some species flattened on top or with a slightly elongate ostiole. The asci are more cylindric and tend to lie parallel to one another, rather than in a fascicle. The interthecial tissue consists of strips of interthecial parenchyma, similating paraphysoids. The ascospores are two-celled, hyaline and more definitely constricted than in *Mycosphaerella*. Many species are parasitic on leaves and stems, and the ascostromata often mature over winter on the dead parts of the host. The conidia are borne in pycnidia similar to the ascostromata, but of the form genus *Phoma* (conidia one-celled) or *Ascochyta* (some conidia at least, two-celled). *Didymella pinodes* causes an anthracnose of peas;

D. applanata a disease of *Rubus*, and *D. lycopersici* a black spot of tomatoes.

Didymosphaeria: The genus *Didymosphaeria* is an example of the arbitrary placement of a genus because of the inadequate information regarding the manner of formation of the interthecial tissues.

Scheinpflug (**1958**) gives the genus as having immersed, globose ascostromata with a short ostiole, sometimes with some clypeate development in the epidermis and with two-celled, brown ascospores. Niessl (**1875**) included species with two-celled, hyaline ascospores in the genus. Saccardo (**1882**) placed the hyaline spored species in the genus *Didymella*, which Winter gave as a subgenus of *Didymosphaeria*. Theissen and Sydow (**1918**) and most other early writers placed the genus in the family Pleosporaceae under the Pseudosphaeriales (i.e., with filiform interthecial strips, usually called paraphysoids). Müller and von Arx placed *Didymosphaeria* in their Pleosporaceae and *Didymella* in their Mycosphaerellaceae. Munk (**1957**) placed *Didymella* in his family Pleosporaceae, along with the "form genus" *Mycosphaerella* and *Pleospora*, and made a new family Didymosphaeriaceae for *Didymosphaeria*. Scheinpflug (**1958**) places *Didymosphaeria* in the family Pleosporaceae with some doubt.

As a matter of fact, the interthecial filaments in both *Didymella* (Corbaz, **1957**) and *Didymosphaeria* (Scheinpflug, **1958**) are identical, and as *Didymella* is usually separated from *Mycosphaerella* in the presence of these filaments and other characters of degree only, all these genera should fall in the same family (Mycosphaerellaceae). We have no detailed evidence that the "paraphysoids" in these genera arise before the origin of the asci.

Keissleriella: This genus is similar to *Didymella* in that it has globose immersed ascostromata, found on herbaceous and woody stems. It differs in the presence of dark colored setae within the ostiole. It has two-celled, hyaline ascospores. The genus *Trichometasphaeria* has similar setae within the ostioles but is given by Munk (**1957**) as having spores one-four septate, and is placed in the family Massarinaceae. Holm (**1957**) places this genus in the Pleosporaceae. Müller and von Arx (**1962**) pointed out that the spores of *Keissleriella* may be two- or many-celled and placed *Trichometasphaeria* as a synonym of *Keissleriella*. *Trichometasphaeria* is placed in the Pleosporaceae in this account.

Microthelia (*Astrosphaeriella*): *Microthelia* has been considered as a lichen associated genus, but similar forms in *Astrosphaeriella* are found not associated with lichens. The ascostromata are hemispheric conic, with a flattened base and occur superficial or strongly erumpent on woody substrata. The ascospores are two-celled and brown and differ from *Didymosphaeria* only in the hemispheric more superficial ascostromata, according to Scheinpflug (**1958**).

Caryospora: The genus *Caryospora* has ascostromata with a flattened base and differs from *Microthelia* in the larger spores (over thirty μ long) which have a false septum in the end of each cell. The species *Caryospora putaminum* occurs on the surface of the stony seeds of the genus *Prunus* (Jeffers, **1940**).

Leptoguignardia: This genus was described by Müller (**1955**) from stems of *Onobrychis*. The ascostromata are similar to those of *Guignardia*, being subepidermal, somewhat flattened and without an ostiole but opening by a pore. The ascospores are three-celled, with the central cell larger than the end cells, and hyaline. The interthecial tissue consists of strips of modified locule parenchyma. It has a conidial state of the form genus *Diplodina*, forming pycnidia similar to the ascostromata and containing two-celled hyaline conidia, with the upper cell narrower and pointed.

Leptosphaerulina: The species of *Leptosphaerulina* have previously been described under the names of *Pleosphaerulina* and *Pseudoplea*. The ascostromata are small (80-150 μ in diameter) with more or less of an ostiolar neck and formed in large numbers in the host tissue. Several species occur as parasites on the leaves of various legumes, and other species as saprophytes or weak parasites on the leaves and stems of many species of plants. The ascostromata may mature on the necrotic spots on the leaves or may continue to eject ascospores throughout the winter and into the spring. The asci are few in number, and broadly saccate. Graham and Luttrell (**1961**) and Bubak (**1909**) figure a thickened apical wall partially penetrated by a pore, surrounded by a ring-like rim. The ascospores of *Leptosphaerulina* are extremely variable. They are often three-septate at first, but extra septa may be inserted, giving rise to three-six-septate spores. A vertical septum is also usually inserted in the central cells. They are hyaline at first but may turn brown in some species. As a result the species are very difficult to delimit. Graham and Luttrell (**1961**) have monographed the genus.

The development of the ascostroma of *L. gaeumannii* has been reported by Müller (**1951**) under the name of *Pleospora gaeumannii* and by Wehmeyer (**1955**) under the name of *Pseudoplea gaeumannii*. The ascostromata originates by the division of a vegetative cell, according to Müller, and by the division of a short hyphae into short multinucleate cells, according to Wehmeyer. In the resulting globose parenchymatic stroma, Müller recognizes two deeply-staining cells. He believes that one of these cells, the "female" ascogone, produces a short hyphal outgrowth, the trichogyne which fuses with the second "male" cell or antheridium.

According to Wehmeyer, the protoplasts of the original initial divide rapidly by cleavage giving rise to irregularly shaped cells. The outer cells of this mass show a regressive type of development and become thick-walled and pigmented to form the wall. The inner cells show a progressive development and several of them become deeply staining and have enlarged nuclei. These cells without any fusion divide into binucleate cells, the ascogonia of most writers. The asci dissolve away the locule parenchyma leaving only unaltered cells and compressed strips.

The genus *Monascostroma* is probably the same as *Leptosphaerulina*. It was described by von Höhnel (**1918**) as having a single ascus, containing two-celled ascospores, hyaline at first, becoming brown. These characters are variable as in *Leptosphaerulina.*

Wettsteinina: The genus *Wettsteinina* is similar to *Leptosphaerulina* (von Höhnel, **1907**). It differs chiefly in the larger ascostromata (120-700 μ in diameter) and the large spores (30-125 μ long) which have a gelatinous envelope. The ascospores are asymmetric and two-celled at first, but additional transverse and also longitudinal septa are later laid down. The species are mostly parasitic on the leaves and stems of plants in colder climates. *Leptosphaerulina* and *Wettsteinina* grade off into *Pyrenophora* and other genera of the Pleosporales which have true paraphysoids.

Sphaerulina: This genus is similar to *Mycosphaerella*, but has hyaline ascospores with more than two cells.

CAPNODIACEAE:

The family Capnodiaceae includes an artificial group of genera based upon the presence of a superficial, dark colored, dematioid, toruloid mycelium which, in many species, is compacted into ribbon-like or rope-like strands. Many of the ascostromata are vertically elongate. They are common in the tropics, where several genera, e.g. *Scorias* and *Capnodium*, occur on the honey dew of insects. The family has been placed in the Perisporiales by Theissen and Sydow (**1917**), the Sphaeriales by Arnaud (**1910**), the Erysiphales by Miller (**1949**), the Pseudosphaeriales by Gaeumann (**1949**) and Müller and von Arx (**1962**), and in the Dothideales by Luttrell (**1955**), because of the developmental studies of Fraser (**1935**). Höhnel (**1918**) and Theissen and Sydow (**1917**) give keys to the genera. Theissen and Sydow include a number of genera which are here placed in the Perisporiopsidaceae.

I. Spores 1-celled .... CAPNODIELLA
II. Spores 2-celled
  A. Spores hyaline
    1. Associated with lichens .... ECHINOTHECIUM
    2. Not associated with lichens
      a. With setae on the mycelium and ascostromata .... CHAETOTHYRINA
      b. Without setae .... CALYPTRA
  B. Spores brown
    1. Asci 8-spored .... LIZONIA
    2. Asci 16-spored .... PSEUDOLIZONIA
III. Spores phragmosporous (with more than 2 cells)
  A. Ascostromata stalked or vertically elongate
    1. Spores hyaline
      a. Mycelium compacted into rope-like strands .... SCORIAS
      b. Mycelium dark, membranous .... ANTENNELLA
    2. Spores brown .... CAPNODARIA
  B. Ascostromata sessile, spherical
    1. Setae present on mycelium or ascostromata
      a. Spores hyaline
        i. Setae in the ostiole only .... AITHALODERMA
        ii. Setae on mycelium or ascostromata .... CHAETOTHYRIUM
      b. Spores brown .... SETELLA
    2. No setae present
      a. Spores hyaline .... LIMACINIA
      b. Spores brown .... PHRAGMOCAPNIAS
IV. Spores muriform (both transverse and vertical septa)
  A. Ascostroma vertically elongate .... CAPNODIUM
  B. Ascostroma spherical superficial
    1. Setae present; spores hyaline .... TREUBIOMYCES
    2. Setae lacking
      a. Spores hyaline to pink to light brown .... PHAEOSACCARDINULA
      b. Spores brown .... COCCODINIUM
V. Spores filiform .... ACTINOCYMBE

The development of *Aithaloderma ferruginea* was studied by Fraser (**1935**). The mycelium forms a dark weft on the host (*Citrus*). Pycnidia first arise from several cells of the mycelium, which divide to form a flat plate of cells. These cells give rise to upward growing hyphae which form the wall of the pycnidium. The pycnidium is wide open, cup-shaped and has bacilliform pycnidiospores budded off from the lower wall cells.

The ascostromata form a similar disc of cells. Hyphae grow out from the under side of this plate as radial growth proceeds at the margin. This growth forms a central dome-shaped mass of parenchyma. An archicarp consisting of several enlarged deeply staining cells, arises in the base of this stroma. This archicarp at first becomes septate and the septa then break down, forming a large multinucleate cell. Pairing of the nuclei and fusion then takes place, apparently, in this enlarged cell. This cell becomes lobed and sends out the ascogenous hyphal system. The ascostroma increases in size to form a spherical ascocarp. The asci arise from buds of the ascogenous hyphae. They contain a single (diploid?) nucleus and show no nuclear fusion in the ascus. The developing asci dissolve away the surrounding parenchyma. An ostiole is formed with dark setae within it. The ascospores are three-five septate, cylindric and hyaline.

Fraser also studied *Capnodium salicinum* which grows on the honey dew excreted by insects. Pycnidia arise by the division of several mycelial cells to form a flat plate. In this species hyphae grow up from this plate to form a bottle-shaped pycnidium. The pycnidiospores cut off from the interior cells are muriform and brown. In the case of the ascostroma, a fascicle of hyphae grow up from a similar plate of cells to form an upright stalk. At the meristematic apex of this stalk a core of small, light colored, loose hyphae arise. The archicarp arises in the base of this core as a multinucleate hypha which branches to form the ascogenous hyphae. The asci, here also, are uninucleate from the first and no nuclear fusion was seen. The ascospores are muriform and brown. The core tissue is entirely absorbed by the developing asci, but Fraser figures a few filiform paraphysis-like hyphae which later grow up between the asci.

ENGLERULACEAE:

The Family Englerulaceae was originally based upon the gelatinous character of the ascostromal tissues at maturity. The upper wall breaks down more or less completely as a slime mass, when the ascospores are discharged. This character, however, has brought together species whose relationships are with many other taxa (Petrak, **1928**). Such gelatinization of the tissues occurs in the Myriangiales (Myxomyriangiaceae), Microthyriales (Myxothyriaceae), Meliolaceae, Hypocreaceae, etc.

Müller and von Arx (**1962**) limit the family to superficial spherical ascostromata similar to those found in the Perisporiopsidaceae, but differing in the gelatinization of the tissues. These fungi are superficial leaf parasites with a superficial mycelium and spherical ascostromata which arise above, rather than beneath the mycelial hyphae, as in the Microthyriales. The asci are saccate to broad clavate with a thickened apical wall and occur in small groups of one-six within the ascostromata. The following key is adapted from that of Müller and von Arx.

I. Mycelium without hyphopodia .... ENGLERULA
II. Mycelium with hyphopodia
   A. Ascostromata sessile on the mycelium
      1. Mycelium with fusoid, 2-3-septate conidia, and globose many-celled conidia . ... SCHIFFNERULA
      2. Mycelium without such conidia
         a. Ascostromata with secondary hyphae growing up around them .... PARENGLERULA
         b. Without such support hyphae .... RHYTIDENGLERULA
   B. Ascostromata on short stalk-like hyphae and with a single ascus ...... THRAUSTE

Englerula: This genus occurs on living leaves, has no hyphopodia on the superficial mycelium, and may form pycnidia containing one-celled brown conidia of the form genus *Capnodiastrum* (*Oothecium*) which differs from *Asteromella*, found in *Asterina*, in being formed on instead of beneath the mycelial hyphae.

Rhytidenglerula: This genus also forms *Capnodiastrum* pycnidia with conidia which may have a light colored equatorial band. Its mycelium has hyphopodia.

Schiffnerula: *Schiffnerula* has conidia similar to those found in *Clypeolella* of the Asterinaceae. In one type, there are formed two-three-septate fusoid conidia, in which the central cells are dark- and the end cells light-colored, and formed on lateral branches of the mycelium (form genus *Mitteriella*). In the other type, many-celled balls of brown cells are borne on short hyphae (form genus *Sarcinella*).

Parenglerula: This genus has several septate support hyphae growing up around the ascostroma, from a basal group of cells.

Thrauste: In *Thrauste* the ascostromata contain only a single large ascus. The ascostroma is borne on a short sterile hypha.

The genera of the family have been discussed by Petrak (**1928**).

PERISPORIOPSIDACEAE:

The Perisporiopsidaceae differ from the Englerulaceae in the lack of gelatinization of the tissues. In their morphologic and ecologic aspect they are similar to the Erysiphaceae, Meliolaceae and Dimeraceae. They occur as superficial leaf parasites with a superficial mycelium, but differ in the ascostromata having only a few saccate or broadly clavate asci with a thickened apical wall and the lack of appendages on the ascocarp. The mycelium, here, may penetrate somewhat more deeply into the leaf tissue. The ascostromata have no true ostiole but form a perforate opening at spore discharge.

Many of the species have characteristic hyphomycetous conidia on the superficial mycelium.

The family is placed in the Dothideales because of the lack of any true paraphysoids, and the saccate, thick-walled asci.

The following key is an adaption of that of Müller and von Arx (1962).

I. Ascostromata without an hypostroma
  A. Spores more than 30 μ long .... PERISPORIOPSIS
  B. Spores less than 30 μ long
    1. Ascostromata fused in a crust on a subiculum................ NEOPARODIA
    2. Ascostromata sessile or on short stalks, with setae or hairs
      a. Mycelium with hyphopodia .... BALLADYNA
      b. Mycelium without hyphopodia .... DYSRHYNCHUS
II. Ascostromata on a basal hypostroma
  A. Spores 30-50 μ long
    1. Ascostromata grouped on a subiculum; spores 2-4-celled .. CHEVALIEROPSIS
    2. Ascostroma with subepidermal hypostroma .... SCOLIONEMA
  B. Spores 18-30 μ long
    1. Ascostroma single or grouped on a basal foot, with setae
      a. Ascostroma on a subcuticular hypostroma .... KUSANOBOTRYS
    2. Ascostroma on a superficial mycelium .... STOMATOGENE
  C. Spores 12-15 μ long.. ALINA

Perisporiopsis: In this genus the ascospores are very large (30-80 μ long), two-celled at first but becoming three-four-celled. Two- to three-celled brown fusoid macroconidia of the form genera *Sporhelminthium, Diploidium* or *Septoidium*, are borne on short side branches of the mycelium. Some species also form lemon-shaped hyaline microconidia on lateral branches of the mycelium.

Scolionema: This genus also has large ascospores (40-65 μ long). The ascostromata are scattered in a superficial subiculum.

Neoparodia: In *Neoparodia* the ascostromata are confluent in a crust on a surface subiculum. Muriform conidia are reported on the mycelium.

Chevalieropsis: This genus forms a superficial basal stroma on which the ascostromata occur in confluent groups. Conidiophores arise from the basal stroma, upon which one-celled microconidia and dark colored, stellate, many-celled macroconidia are produced.

Kusanobotrys: This forms a subepidermal hypostroma, from which erumpent, tuberculate, setose stromata arise. The ascostromata are formed in clusters on this stromatic foot.

Stomatogene: In this genus the ascostromata also occur singly or in small groups upon a basal stalk or foot, which penetrates the stoma. The ascospores are two-celled and brown.

Alina: This genus has very small ascospores (12-15 μ long). There is a superficial dematioid mycelium, upon which there are flask-shaped conidiophores which bear long fusiform, two-six-celled dark colored conidia of the form genus *Sporidesmium*.

Balladyna: This genus has *Clasterosporium*-like conidia, which are long clavate, straight or curved, five-ten-septate and brown, occurring on the mycelium of *B. vandersytii*. The species of this genus have a surface mycelium with hyphopodia, and may have setae. The ascostromata are small and sessile or borne on short stalk-like side branches of the mycelium.

Dysrhynchis: This differs from *Balladyna* only in the lack of any hyphopodia on the superficial mycelium. *Dysrhynchis pulchella* is reported as having a *Coremium* type conidial state. The coremia have a bulbous base and an elongated stalk of parallel hyphae which flare out at the apex into branched conidiophores which bear elongate (15-28 μ) one-four-celled brown conidia.

PLEOSPORALES

The family Pleosporaceae was placed by Lindau (**1897**), Winter (**1887**), Ellis and Everhart (**1892**) and other earlier writers, in the Sphaeriales, and included species with immersed "perithecia" containing paraphyses. With the erection of von Höhnel's Pseudosphaeriaceae, many of these fungi were transferred to that family which was raised to ordinal rank as the Pseudosphaeriales by Theissen and Sydow (**1917**). Later, Luttrell (**1955**) pointed out that the total dissolution of the locule parenchyma, the unaltered remains of such parenchyma or the compression of this parenchyma into interthecial strips or filaments, were merely matters of degree in the modification of the locule parenchyma. Inasmuch as the type genus, *Wettsteinina*, of the Pseudosphaeriales, contains unaltered locule parenchyma at maturity, Luttrell placed the name Pseudosphaeriales as a synonym of the Dothideales. He then followed Miller (**1938, 1949**) in recognizing that there was another distinct type of interthecial tissue, in which the centrum parenchyma of the locule elongated by intercalary growth into filaments, before the asci arose, which he called pseudoparaphyses and which I term paraphysoids in the present work. He used the ordinal name Pleosporales for this group, and placed the families Pleosporaceae, Venturiaceae and Lophiostomataceae in this order. Although this type of paraphysoid formation has been demonstrated for several genera in this order, the great majority of genera have not been studied in this detail. Genera and even families of the Loculoascomycetes must therefore be placed in the order arbitrarily.

Müller and von Arx used the ordinal name Pseudosphaeriales to include all the Loculoascomycetes with true ostioles, and the Pleosporaceae for a group of genera with "paraphysoids."

Since the terms paraphyses and paraphysoids have been loosely used for any filamentous interthecial hyphae, it is difficult to place genera from the literature according to this family and ordinal differentiation.

PLEOSPORACEAE:

The family Pleosporaceae has never been clearly delimited under the modern concept of Luttrell, as those fungi with true paraphysoids, consisting of elongated hyphal threads in the centrum of the locules, formed before the asci arise. Many members of the Loculoascomycetes have not been studied as to the manner of formation of such interthecial filaments. Therefore, many genera which have no other family affiliations must be placed here arbitrarily. This is apparent from the didymosporous genera placed in this family by Müller and von Arx. These genera are a mixed lot with many different types of ascostroma development. They are included in the following key, however, with a discussion of their probable affinities under the respective genera. As a result, the definition of the family Pleosporaceae is not clear and ascostromata of many different types, which were previously placed in various other families (Trichospheriaceae, Melanommeaceae, Cucurbitariaceae, Amphisphaeriaceae etc.) are to be found here.

KEY TO THE GENERA

I. Spores 2-celled (many-celled in some genera)
  A. Ascostromata superficial
    1. Ascostromata fleshy, light colored on a short stalk or hypostroma
      a. Ascostromata tomentose; spores hyaline . . . . LETENDRAEA
      b. Ascostroma smooth; spores gray-brown . . . . ALLONECTE
    2. Ascostromata dark colored
      a. Ascostroma setose, tomentose or on a subiculum
        i. Ascostroma setose. . . . . HERPOTRICHIELLA
        ii. Ascostroma tomentose or on a subiculum . . . . HERPOTRICHIA
      b. Ascostromata smooth, not on a subiculum
        i. Spores brown, parasites on legumes; ascostromata entirely superficial, in series. . . . . PARODIELLA
        ii. Spores hyaline; ascostromata with flattened base partially immersed; saprophytes on wood . . . . MELANOPSAMMA
  B. Ascostromata subcuticular
    1. Ascostromata shield-shaped with a tomentose margin . MYCOMICROTHELIA
    2. Ascostromata grouped under a superficial crust
      a. Ascostromata separate, under a superficial crust with a tomentose margin; associated with lichens . . . . TOMASELLIA

b. Ascostromata multilocular around a sterile center; not tomentose....... .... GILLETIELLA

C. Ascostromata more deeply seated in the host tissue
   1. Hyperparasites in rust pustules or other fungi................. EUDARLUCA
   2. Parasites or saprophytes
      a. Spores hyaline, with appendages at the septum (marine).. PARALIOMYCES
      b. Spores brown or colored
         i. Spores long cylindric; ascostromata in concentric rings on leaf spots.. .... TERATOSPHAERIA
         ii. Spores ellipsoid, often apiosporous
            (a). Spores large, 23-95 μ long, with a gelatinous envelope........... .... PTERIDIOSPORA
            (b). Spores less than 25 μ long, without a gelatinous envelope .... DIDYMOPLEELLA

D. Ascostromata unilocular, confluent or grouped on a basal stroma
   1. Ascostromata erumpent, smooth, on a basal stroma .... OTTHIA
   2. Ascostromata superficial in a felt of mycelium.................. LICOPOLIA

II. Spores phragmosporous (many-celled) when mature (2-celled in some species or when immature)

A. Ascostromata superficial, spores brown
   1. Ascostroma tomentose or on a subiculum .... HERPOTRICHIA
   2. Ascostroma setose.................................. HERPOTRICHIELLA
   3. Ascostroma smooth
      a. Ascostromata on an erumpent basal hypostroma........... GIBBERIDEA
      b. Ascostromata not on an erumpent hypostroma
         i. Base of ascostromata flattened, slightly immersed in the substratum; spores large, 25-90 μ long.................. TREMATOSPHAERIA
         ii. Superficial; spores smaller, less than 25 μ long
            (a). Spores brown .... MELANOMMA
            (b). Spores hyaline................................. ZIGNOELLA

B. Ascostromata immersed
   1. Spores hyaline
      a. With setae in the ostiole .... TRICHOMETASPHAERIA
      b. Without setae in the ostiole .... METASPHAERIA
   2. Spores colored (pale yellow to dark brown)
      a. Ascostromata scattered; singly erumpent .... LEPTOSPHAERIA
      b. Ascostromata scattered, confluent or in valsoid groups, with some secondary mycelial development around them.................. THYRIDARIA

III. Spores muriform (with both transverse and vertical septa)

A. Ascostromata superficial.................................. TEICHOSPORA

B. Perithecia clustered, partially immersed, erumpent, in a secondary mycelial stroma or on a basal stroma
   1. Unilocular ascostroma on a basal stroma
      i. Hypostroma dark colored, parenchymatic.............. CUCURBITARIA
      ii. Hypostroma light colored, scleroplectenchymatic ...... CUCURBIDOTHIS
   2. Ascostroma scattered, confluent or in an erumpent stoma, or valsoid; spores small, dark brown....................................... THYRIDIUM
   3. Ascostromata in a valsoid secondary stroma; ostioles collectively erumpent; spores larger, yellow-brown.............................. FENESTELLA

C. Ascostromata immersed and separately erumpent
   1. Spores colored
      a. Ascospores flattened in one plane
         i. Spores with 1 (rarely 2) vertical septa .... PLATYSPORA
         ii. Spores with 3-7 vertical septa .... CLATHROSPORA

b. Ascospores not flattened in one plane
   i. Ascostromata large, 135-450 μ in diameter; spores 35-100 × 10-40 μ, not more than 7-septate; ascostromata often with stout setae; paraphysoids poorly developed .... PYRENOPHORA
   ii. Spores 9-70 × 5-25 μ, 2-14 septate; ascostroma smooth, tomentose or setose; paraphysoids well developed .... PLEOSPORA
   iii. Spores with a gelatinous envelope .... PLEOMASSARIA
2. Spores hyaline .... CATHARINIA

IV. Spores scolecosporous
   A. Ascospores with deep constriction in middle, with a swollen cell above and below constriction .... OPHIOBOLUS
   B. Ascospores of uniform diameter; lacking deep constriction and swollen cells .... COCHLIOBOLUS

There is a central group of genera, including *Leptosphaeria, Trichometasphaeria, Pyrenophora, Pleospora, Ophiobolus, Cucurbitaria* and others, in which the presence of true paraphysoids has been demonstrated. These genera represent a nucleus of the true Pleosporaceae. A large number of conidial connections have been obtained in culture of these genera by Drechsler (**1923, 1934**). Lucas (**1963, 1965**), Luttrell (**1963,1964**), Riedl (**1959**), Webster (**1951, 1955, 1955a, 1957**), Webster and Hudson (**1957**), Webster and Lucas (**1959, 1961**), and Lucas and Webster (**1964, 1967**), and many others. An examination of the following list of such connections gives an opportunity to weigh the importance of the connection of the perfect and imperfect states of such fungi. There is no absolute correlation between the form genera and the ascigerous genera. Certain considerations, however, must be taken into account in rating any form genera in this list. It can be seen that both hyphomycetous and pycnidial forms occur. If one takes into consideration the spore form, septation and color, regardless of whether they occur in pycnidia or on free conidiophores, the correlation is more apparent. Many of the pycnidial states, such as *Phoma, Leptophoma* and *Plenodomus*, have been considered by some writers as spermagonia or microconidia. In fact, Shoemaker (**1955**) has shown that such spermagonia function in fertilization between the sexual strains of some heterothallic species. In *Leptosphaeria* the pycnidial forms have septate and for the most part phragmosporous conidia. Only *Coniothyrium* has one-celled colored conidia. *Diplodina* (spores two-celled, hyaline), *Stagonospora* (spores many-celled, hyaline), *Hendersonia* (spores many-celled, colored), and *Septoria* (spores filiform, colored), are all phragmosporous as are the ascospores. The hyphomycetous form genera *Vakrabeeja* and *Rhabdospora* also have phragmosporous conidia. A few species of *Leptosphaeria* have muriform conidia in the form genus *Alternaria*. In *Pleospora* there are similar pycnidial states reported, but there is a tendency to produce muriform conidia of the form genera *Alternaria, Stemphylium* and *Camarosporium*, corresponding to the muriform ascospores. *Ophiobolus* and *Cochliobolus* again have the phragmosporous conidia of *Septoria* and *Helminthosporium*, as in their ascospores. In *Cucurbitaria* again are muriform ascospores and conidia as in the form genus *Camarosporium*, with *Phoma* also present. *Trichometasphaeria* has the phragmosporous *Helminthosporium* conidia.

The following are some of the cultural connections made between the perfect and the imperfect states, as reported in the literature.

Keissleriella cladophila	*Dendrophoma*
Trichometasphaeria turcica	*Helminthosporium turcicum*
Trichometasphaeria pedicellatum	*Helminthosporium pedicellatum*
Trichometasphaeria holmii	*Helminthosporium holmii*
Trichometasphaeria gallica	*Ascochyta*
Leptosphaeria agnita, L. doliolum (*Plenodomus*), *L. purpurea* (*P. sanguiolenta*), *L. solani, L. spartinae, L. maculans,* (*Plenodomus rabenhorstii*), *L. submaculans, L. typhicola, L. haematites, L. libanotis*	*Phoma*
L. acuta	*Leptophoma*
L. artemisiae	*Asteromella*
L. bondari, L. michotii (*C. scirpi*), *L. obtusispora* (*C. henriquesii*), *L. faullii* (*C. faullii*), *L. periclymeni*	*Coniothyrium*
L. typharum, L. vitalbae, L. viridella	*Hendersonia*
L. dumetorum, L. pontiformis, L. libanotis, (*Hendersoniella*)	*Hendersonia* and *Phoma*
L. marcyensis	*Diplodina*
L. pratensis, L. bondari	*Stagonospora* and *Phoma*
L. vagans (var. *shepherdiae*), *L. eustoma, L. fuckelii* (*S. bromi*), *L. nigrans, L. eustomoides, L. nodosum* (*S. nodosum*)	*Septoria*
L. microscopica (*P. festucae*), *L. macrospora*	*Phaeoseptoria*
L. heterospora, L. culminfraga, L. modesta	*Alternaria*
L. salvinii	*Vakrabeega*
L. anemones	*Rhabdospora*
Pleospora rubicunda, P. betae	*Phoma*
P. cytisi	*Leptophoma*
P. spartium-juncei, P. shepherdiae	*Coniothyrium*

P. inverecunda	*Staganospora opuntiae*
P. clavescens	*Microdiplodia henningsii*
P. vagans	*Hendersonia crastophila*
P. media	*Camarosporium*
P. valesiaca, P. alternariae	*Alternaria*
P. herbarum	*Stemphylium botryosum*
P. orbicularis, P. flavo-fusca (*Stemphylium?*), *P. scrophulariae*	*Alternaria* and *Phoma*
Pyrenophora typhaecola	*Phoma*
P. trichostoma	*Helminthosporium teres*
P. scirpi (*Pleospora ?*)	*Alternaria scirpicola*
Pyrenophora spp.	*Helminthosporium*
Clathropsora elynae, C. diplospora	*Alternaria*
Ophiobolus herpotrichus	*Septoria* (*Urohendersoniella*)
O. (*Cochliobolus*) *heterostrophus*	*Helminthosporium maydis*
Cochliobolus sativus	*Helminthosporium sorokinianum* (*H. sativum*)
C. spicifera	*Helminthosporium spiciferum*
C. carbonum	*Helminthosporium carbonum*
C. victoriae	*Helminthosporium victoriae* and *Phoma*
C. lunatus	*Curvularia lunata*
C. intermedia	*Curvularia intermedia*
Cochliobolus spp.	*Euhelminthosporium* (*Bipolaris*)
Cucurbitaria elongata (*C. robinieae*), *C. coluteae* (*C. coluteae*, Riedl, **1959**), *C. spartii* (*C. coluteae*, Müller and Baumeister, **1957**)	*Camarosporium*
C. caraganae	*Camarasporium caraganae* and *Phoma*
C. berberidis	*Pyrenochaeta* (*Phoma*) *berberidis*
C. staphula	*Pseudodichomera*
C. laburni	*Camarasporium laburni*

The genera *Didymosphaeria, Astrosphaeriella, Keissleriella* and *Caryospora*, which Müller and von Arx place in their Pleosporaceae, have here been placed in the Mycosphaerellaceae because of their affinities with *Didymella*. If the sterile filaments in these genera prove to be true paraphysoids, they should be replaced in the Pleosporaceae.

Letendraea and *Allonecte* were originally placed in the Hypocreales because of their light colored ascocarps. Müller and von Arx state that they have bitunicate asci.

Herpotrichia: The genus *Herpotrichia* was placed in the Trichosphaeriaceae in the older classifications. Müller and von Arx give *Neopeckia* (two-celled ascospores) and *Pseudotrichia* (hyperparasites in pyrenomycetous ascostromata) as synonyms of *Herpotrichia*. The ascostromata of *Herpotrichia* are superficial and either on a mycelial subiculum or with tomentose walls. *Herpotrichia* (*Neopeckia*) *coulteri*, occurring on pines and *H. juniperi* (*H. nigra*), occurring on various conifers, cause a snow blight of the needles and grow beneath the snow where the temperature and moisture conditions are favorable. The mycelium grows out over the surface of the needles and mats them together with a thick black mass of hyphae. The ascostromata occur in or on the surface of this subiculum. The ascospores are brown and may be two-celled (*Neopeckia*) or many-celled (*Herpotrichia*). Other species of *Herpotrichia* occur on dead and decaying wood. *Herpotrichia* (*Pseudotrichia, Khekia*) *mutabilis* is found in or on the ascostromata of other pyrenomycetes. The ascostromata are covered with short bristle-like setae.

Parodiella: In the genus *Parodiella* the ascostromata are globose, smooth and are clustered as parasites on various genera of the legumes. The ascospores are two-celled and brown with striate walls. Tilak (**1963**) described and figured true pseudoparaphyses which grow downward from the top wall cells. In some Indian species of *Parodiella* smaller astomous pycnidia are also formed.

Gilletiella, Mycomicrothelia and *Tomasellia* have flattened ascostromata, which suggest the Microthyriales.

Eudarluca: This genus occurs as a parasite in the pustules of the rusts and fruit bodies of other fungi. The spores are two-celled and hyaline at first but may become four-celled. A pycnidial form, *Darluca filum*, occurs with two-celled hyaline pycnidiospores bearing a short appendage at both ends.

Paraliomyces: This is a marine genus with two-celled, hyaline ascospores which have a disc-shaped appendage at the septum.

Teratosphaeria: This is a leaf parasite with separate ellipsoid, immersed ascostromata, grouped in the leaf tissue. The ascospores are elongate cylindric, two-celled, brown and echinulate. Pycnidia of the form genus *Dendrophoma* are associated with it. The pycnidiospores are formed on long filiform conidiophores.

Pteridiospora: This genus has large immersed ascostromata surrounded by mycelial growth. The spores are large, brown, apiosporous in many species, and are surrounded by a gelatinous envelope. It is related to the Massariaceae (*Massariella*) as originally delimited, but does not belong to the *Massaria inquinans* group.

Didymopleella: This genus has immersed ascostromata similar to those of *Didymosphaeria*, but with apiosporous ascospores which are unequally two-celled, brown and with the smaller cell lighter in color.

Melanopsamma: This genus was placed in the Melanommeae of the older classifications. It is placed as a synonym of *Chaetosphaeria* under the Sphaeriaceae by Müller and von Arx. The ascostromata are superficial or with a slightly immersed base and the ascospores are two-celled and hyaline. Further information is needed for its proper placement.

Otthia: The genus *Otthia* was placed in the family Cucurbitariaceae in the older arrangements. It has unilocular ascostromata which are clustered or confluent on a basal parenchymatic stroma. The spores are two-celled, brown or hyaline. The hyaline-spored species have been placed in a separate genus, *Otthiella*.

Licopolia: This genus also has unilocular ascostromata scattered on a basal parenchymatic stroma, from which a mycelial subiculum arises about the fruiting bodies. The ascospores are two-celled and light brown.

Melanomma: The genus *Melanomma* was originally placed in the family Melanommeae. Its ascostromata are superficial, somewhat carbonaceous and often clustered or confluent, occurring on dead wood. The ascospores are fusoid, four-celled, brown and with the upper half somewhat larger than the lower half. The development of *M. pulvis-pyrius* has been studied by Chesters (**1938**). Pycnidia of the form genus *Aposphaeria* were formed in culture and occasionally occur in nature. They are immersed in a coarse brown mycelium, which also envelopes the young ascostromata. The pycnidiospores are ovoid, hyaline and are formed endogenously from phialids in a manner characteristic for such conidiophores. The ascostromata arise by the division of vegetative cells to form a parenchymatic stroma. As this stroma increases in size, the central cells become deeply staining. They separate and elongate to form what Chesters calls trabeculae. These trabeculae branch and anastomose and are attached at both the top and bottom to the wall cells. They arise before the asci appear and are typical paraphysoids. The asci grow up between these paraphysoids but do not entirely disintegrate them.

Gibberidea: Holm (**1968**) limits the genus *Gibberidea* to the single type species *G. visci*, which has ascospores which are cylindric, yellow-brown, and are divided at maturity into two four-septate halves by a central constriction. Pycnidia which appear similar to the ascostromata have been placed in the form genus *Pleurostromella*. *Cucurbidothis* (Holm, **1967**) and *Rosenscheldia* differ in the scleroplectenchymatic hypostroma. *Cucurbidothis* has ascospores which are brown and may be either phragmosporous or dictyosporous; in *Rosenscheldia* they are long and spindle-shaped, two-septate, and pale in color. These genera are related to *Leptosphaeria* according to Holm (**1968**).

Thyridaria: The genus *Thyridaria* is closely related to *Melanomma*. It differs in the immersed ascostromata, which usually occur in an irregular or valsoid group, and are surrounded by a felty growth of mycelium. The valsoid species are erumpent through a common disc, but the ostioles are separately erumpent. A few species which have scattered separately erumpent ascostromata have been placed in a separate genus, *Kalmusia*. The ascospores are four-celled, as in *Melanomma*. The genus has been monographed by Wehmeyer (**1941b**). Chesters (**1938**) has described the development in *Thyridaria rubro-notata.* Both ascocarps and pycnidia arise in a felty stroma of reddish mycelium. The pycnidia are similar to those of *Melanomma* but the slimy phialospores are colored rather than hyaline. The ascocarps develop as in *Melanomma*. Filamentous, branching trabeculae, as in *Melanomma*, are formed. The ascostromata are covered with a reddish powder.

Thyridium: Thyridium differs from *Thyridaria* only in the muriform character of the ascospores. The three genera *Melanomma, Thyridaria* and *Thyridium* compose a group which could be the basis of a separate family (Melanommaceae).

Trematosphaeria: This genus differs from *Melanomma* in the large, more hemisphaeric or conic ascostromata with the base partially sunken in the substratum. The ascospores are much larger (25-95 μ long) than in *Melanomma*, and in most species have more than four cells. The genus was formerly placed in the Amphisphaeriaceae.

Leptosphaeria: Leptosphaeria is a very large genus with several hundred described species. The ascostromata are immersed within the substratum and the ascospores are three- to many-septate, pale yellow, olive to dark brown with various types of septation. They occur as parasites or saprophytes on leaves and stems of higher plants. In many species the ascostromata mature saprophytically on the dead plant parts. The species of Switzerland have been monographed by Müller (**1950**), and those of Sweden by Holm (**1952**). Some species are limited to certain genera of hosts, while others are ubiquitous. Holm, in his study of the Pleosporaceae (**1957**), divided the genus into several genera, but the outlines of these genera are rather vague. They are as follows:

Leptosphaeria is limited to species related to *L. doliolum.* These have an inner wall layer of "scleroplectenchyma" which he defines as a tissue composed of rounded, thick-walled, hyaline cells. However, he gives the walls of some species of *Nodulosphaeria* and *Phaeosphaeria* as also scleroparenchymatic.

Nodulosphaeria, according to Holm, includes species which have pale olive ascospores which have an inflated cell in the upper half of the spore, and in most species have small hyaline apical appendages. However, spores with inflated cells are also found in *Leptosphaeria* and *Phaeosphaeria.*

Phaeosphaeria: This genus has the ascostromata often arranged in series. The walls are thin and composed of small cells. The ascospores are yellow-olive and with inflated cells, but no appendages.

Entodesmium: This genus has ascostromata with elongated necks, containing paraphyses, and thin walls of small cells. The asci are numerous and the ascospores fragment in some species.

Leptospora: This genus is limited to a single species with filiform spores, elongated ostiole and causing a reddish discoloration of the substrate. The species was formerly placed in *Ophiobolus*.

Ophiobolus: This genus is here limited to two species which have filiform, many-sepatate ascospores in which two central cells are inflated. The ascospores tend to fragment between the two inflated cells. Holm placed the other species with filiform spores in the various genera mentioned above.

Leptosphaeria bondari, with *Coniothyrium* and *Stagonospora* as conidial states, causes an aureolate leaf spot of *Citrus* in the tropics (Bitancourt and Jenkins, **1935**). *Leptosphaeria maculans* with the conidial state of *Plenodomus* (*Phoma*) *maculans,* causes a leaf spot, dry rot or black leg of various species of *Brassica* (turnip, broccoli, cabbage etc.).

The development of *Leptosphaeria cavanillesii* was reported by Urries (**1946**). The ascostroma here originates by the division of cells of the vegetative mycelium. A central group of deep staining cells differentiate and one of them becomes binucleate. This binucleate cell divides to form a number of binucleate cells which in turn give rise to the ascogenous hyphae. Meanwhile there is an elongation and division of the central cells of the locule to form the sterile paraphysoids. The asci then grow up between these paraphysoids. This is a typical Pleosporaceaeous development.

Species of *Leptosphaeria* are found on *Juncus, Spartina* and *Phragmites* in marine habitats.

The genus *Metasphaeria* was formerly used for those species which are similar to *Leptosphaeria* but had hyaline many-celled spores. Most of these species have been transferred to other genera.

Trichometasphaeria: This genus is based on the occurrence of setae in the ostiole and its many-celled, hyaline ascospores. *Keissleriella* is a similar genus, but with two-celled ascospores. *Trichometasphaeria turcica* and *T. holmii* are two closely related species, with conidial states of *Helminthosporium* (*Bipolaris*) *turcicum* and *H. holmii* respectively, which have been studied by Luttrell (**1958, 1963**). The ascostromata of *T. turcica* arise as small spherical masses of parenchyma, by the division of vegetative cells. Enlarged, multinucleate, deeply staining ascogonial cells soon appear in these stromata. Protruding hyphae, which could be called trichogynes were seen but no actual fusion between plus and minus strains was observed. The ascogenous cell gave rise to ascogenous hyphae, upon which binucleate cells appeared. Asci were formed by the development of croziers. Before the asci arose, the central cells of the locule parenchyma began to elongate downward to form a branching mass of filaments. The asci grew upwards through these paraphysoids. This is a typical Pleosporaceous development. The species of *Trichometasphaeria* are heterothallic (Luttrell, **1963, 1964**; Nelson, **1965**).

Ophiobolus: The genus *Ophiobolus* includes species with long filiform ascospores, which are many-septate and fragmenting in some species. Some species have inflated cells, and the genus is limited by Holm (**1952**) to the two species which have two central cells inflated in the ascospore, and may fragment between these two cells. He places one species, *O. rubella*, which causes a reddish discoloration of the substratum, in the genus *Leptospora*. The Swedish species of *Ophiobolus*, have been monographed by Holm (**1948**) and the Swiss species by Müller (**1952**). For many years, one of the pathogens causing take-all, an important disease of the roots of wheat, was called *Ophiobolus graminis*. However, Holm (**1948**) and von Arx and Olivier (**1952**) have pointed out that the asci of this species are thin-walled, unitunicate and have a refractive ring in the apex, as is found in the Diaporthales. Von Arx and Olivier have placed it in a new genus *Gaeumannomyces*.

Cochliobolus: The genus *Cochliobolus* was separated from *Ophiobolus* by Drechsler (**1934**) on the basis of the ascospores, which are spirally coiled in this genus. *Cochliobolus sativus* has been studied by Tinline (**1951, 1954**), Tinline and Dickson (**1958**), Shoemaker (**1955**) and Hrushovitz (**1956**). The conidial state is *Helminthosporium* (*Bipolaris*) *sativum*. The species is heterothallic. Spermagonia occur on all strains as large stromata within which minute globose spermatia are cut out endogenously from phialids, in a central slime. Archicarps arise directly on the vegetative mycelium. These consist of an enlarged, multinucleate basal cell and an elongated trichogyne. Additional trichogynes arise from the basal cell, or ascognonium. After three-seven days, spherical stromatic initials containing such branched ascogonial cells with trichogynes which protude to the exterior, arise. Conidiophores and conidia also are found on these "protothecia." These "protothecia" can enlarge, but can not mature to the stage of forming ascogenous hyphae unless they are fertilized by spermatia of the opposite compatible strain. Spermatia have been seen attached to the trichogynes. Protothecia do not mature in single spore strains. Protothecia and spermagonia are produced on all strains. The first indication of differentiation is the appearance of a mass of deep staining meristematic cells just below the upper wall. This is the beginning of ostiole formation. These cells produce downward growing filamentous pseudoparaphyses which fill the cavity of the locule. If fertilization occurs, ascogenous hyphae then appear from the ascogonial cells in the base of the locule. Croziers are formed and from them asci arise and grow upward between the pseudoparaphyses. Shoemaker obtained an albino mutation of *C. sativus*. In crosses with the spermagonia of the wild strain, it was shown that all of the cells of the protothecium including the ascogenous cells were homothallic for the female parent, that is all such cells gave rise to hyaline conidia in culture. Not until ascogenous hyphae appeared with paired nuclei, did heterokaryosis appear.

Nelson (**1957, 1959, 1959a, 1960, 1961, 1965**) has found a number of species of *Cochliobolus* to be heterothallic. He has succeeded in obtaining fertile crosses between compatible strains of different species.

Pyrenophora: The genus *Pyrenophora* is intermediate between *Wettsteinina* and *Pleospora*. It is difficult to distinguish this genus on any one character, but there is a group of characters which set it apart. The ascostromata are large (200-700 μ in diameter), have thick stromatic walls, and in most species bear stiff setae which may function as conidiophores and bear conidia. The asci are broad clavate, with a thickened apical wall. The spores are large (30-103 × 13-40 μ), oblong and pale yellow to yellow-brown. The ascostromata occur on the culms of grasses and cereal grains. Its occurrence on different host genera has been used to separate species (*P. bromi, P. teres, P. avenae, P. graminea* etc.) by Diedicke (**1904**) and as physiologic forms by Noack (**1905**).

The form genus *Helminthosporium*, occurring as a conidial state in this genus and also in *Trichometasphaeria, Ophiobolus* and *Cochliobolus*, causes spot and stripe diseases of the leaves of grasses and cereal grains. It is a complex genus and has been studied by Drechsler (**1923-1928**), Luttrell (**1963a, 1964**) and others.

The development of *P. trichostoma* has been studied by Wehmeyer (**1954**). The ascostroma arises from a vegetative hypha by the enlargement and division of its cells, resulting in a tight knot of thick-walled cells. The outer "wall" cells remain thick-walled and pigmented, but the central cells divide rapidly to a hyaline parenchyma of thin-walled cells. Some stromata do not develop asci but remain as sclerotia. In maturing ascostromata some of these central cells degenerate and become empty. Others become deeply staining, nuclei divide and the protoplast becomes lobed and grows out as branched hyphae. These hyphae grow out in an irregular radiate manner into the center of the locule and form the paraphysoids. Certain other cells show a dense protoplasm and larger dividing nuclei. These grow out along the base of the cavity to form the ascogenous hyphae. The asci arise as swollen tips of these ascogenous hyphae. The young ascus is binucleate and nuclear fusion soon takes place. The paraphysoids, here, are not as definite and not as numerous as in *Pleospora.* Spine-like setae grow out from the superficial thick-walled wall cells. It took 144 days for mature ascospores to form.

Pleospora: The genus *Pleospora* is a large one. It is widely distributed, but is most abundant in subalpine and subarctic regions, where the ascostromata mature under the snow over winter in a saprophytic environment. The ascostromata are immersed and then erumpent, on the stems and leaves of many higher plants. Many species are ubiquitous and occur on many different genera. Others are more limited in their distribution, not always on single genera but on certain plant groups as grasses, sedges or dicotyledons. The asci are clavate and have a thickened apical wall. The spores vary greatly in size, shape, septation and color, but they are all muriform, having both transverse and vertical septa. The genus and its segregates have been monographed by Wehmeyer (**1961**).

The development of the ascostroma in *P. armeriae* was published by Wehmeyer (**1955**). The ascostroma arises as a short chain of cells. These cells divide and branch to form a complex knot. Certain cells near the apex of the ascostroma become loosened and begin to elongate. They finally form a palisade of filaments within the center of the locule, the paraphysoids. At this stage some deeply staining cells, the ascogenous cells, appear. They branch and paired nuclei appear in the ascogenous hyphae. The ascus arises as an outgrowth of these hyphae, into which two nuclei migrate and fuse.

Platyspora: In *Platyspora* the ascospores are flattened in one plane. In face view they are fusoid or clavate and three, four, or five-septate, depending upon the species, and one vertical septum runs through the central cells but not the end cells. In edge view the ascospores are fusoid or cylindric, with transverse but no vertical septa. *Platyspora permunda* occurs on many dicotyledonous genera; *P. pentamera* and *P. planispora* are found on both dicotyledons and monocotyledons.

Clathrospora: The genus *Clathrospora* also has flattened ascospores, but they are broadly fusoid and have from five to fourteen transverse septa and two to five longitudinal vertical septa. In edge view, the ascospores are fusoid to cylindric with transverse but no vertical septa (except in *C. multiseptata* which has faint vertical septa). The species of the genus are confined almost entirely to grasses, sedges and *Juncus*.

Teichospora: The genus *Teichospora* is similar to *Pleospora*, but the ascostromata are superficial and occur mostly on decorticated wood. There are, however, certain differences which are correlated with this habit. The ascostromata are usually larger, thicker walled and with more numerous asci and paraphysoids, than in *Pleospora*. The asci are more cylindric and the spores are usually uniseriate and their septation is more variable than in *Pleospora*. On the other hand, there are species of *Pleospora* which occur on woody stems and are erumpent through the periderm. Such species show the above mentioned characters and are intermediate between the two genera. They have been placed in a subgenus, *Teichosporoides* by Wehmeyer (**1961**).

Cucurbitaria: In *Cucurbitaria* the locules are borne singly in ellipsoid stromatic protrusions, clustered superficially upon a basal stroma. The species are mostly wood inhabiting and are erumpent through the periderm. The genus has been monographed by Welch (**1926**). The spores are muriform and brown. The conidial states are in the form genera *Camarosporium* and *Pseudodichomera*, which is a stromatic form of *Camarosporium*. The conidia are also muriform and brown (Müller and Baumeister, **1957**).

Cucurbitaria staphula is found on galls of the branches of *Populus balsamifera* (Arnold and Russell, **1960**). It is associated with *Macrophoma tumefaciens* on these galls. The *Macrophoma* is supposed to be the cause of gall formation and the *Cucurbitaria* may be a secondary invader. Zalasky (**1964**) obtained a *Pseudodichomera* and two types of microconidia in culture of *C. staphula*.

Fenestella: In *Fenestella* there are several ascostromata, similar to those of *Pleospora*, imbedded in a secondary mycelial stroma, in a valsoid manner. The genus was formerly placed in the Melanconideae or the Valsaceae, but the ascostromata are stromatic in character and the so-called stroma is a secondary development. The asci have been shown to be bitunicate. The spores are muriform and brown.

Catharinia: This is a poorly known genus. It is supposed to represent a *Pleospora* with hyaline spores, but it is not closely related to that genus. The ascospores usually have numerous septa. The genus is southern in its distribution.

VENTURIACEAE:

The relationship of a group of genera similar to *Venturia-Gibbera* has been recognized by several workers. These fungi have been studied and published under the name of the Venturiaceae by von Arx (**1952**) and Müller and von Arx (**1950, 1962**). The family characterized by a group of characters. Many of the species have setose ascostromata. The ascospores are hyaline at first but become yellowish, greenish, gray-green to olive-brown when mature. They are two-celled and in many species are apiosporous with the septum below or above the middle. The ascostromata are of many different types, unilocular or plurilocular, immersed or erumpent, superficial, with a subiculum or hypostroma, with or without stromatic development. The only constant character in the family is the color of the ascospores. Practically all the species are parasites on the leaves of higher plants. As a result the genera included here have been taken from many other previously described families. The actual origin of the ascostroma, or the interthecial tissues are known in only a few species and they are, therefore, not easy to fit into the Luttrell arrangement. Some genera have a coarse parenchymatic stroma as in the Dothideaceae and have been placed in that family also.

The following key is an adaptation of the key to genera by Müller and von Arx (**1962**).

I. Ascostromata unilocular, borne superficially, subcuticularly or immersed
  A. Ascostromata in or on a superficial subiculum
    1. Ascostroma opening widely as in the Discomycetes .... PSEUDOPARODIA
    2. Ascostroma with a pore-like ostiole
      a. Ascospores with the septum at or above the middle
        i. Ascostromata setose
          (a). Subiculum of loose hyphae, mycelium or ascostromata setose .... ACANTHARIA
        ii. Ascostromata tomentose or glabrous .... ANTENNULARIA
      b. Ascospores with septum below the middle; ascostromata smooth or with smal peg-like spines................................ APIOSPORINA
  B. Ascostromata not on a mycelial subiculum
    1. Ascostromata borne singly, immersed in the host tissue or erumpent VENTURIA
    2. Ascostromata clustered, subepidermal .... STIGMATEA
    3. Ascostromata borne singly, superficial or subcuticular
      a. Ascospores septate in the lower portion of the spore .... BOTRYOSTROMA
      b. Ascospores septate in the middle or above the middle
        i. With a superficial mycelium....................... SEYNESIELLA
        ii. Without a superficial mycelium .... COLEROA
    4. Ascostromata borne on a stalked hypostroma .... PHAEOCRYPTOPUS
II. Ascostromata plurilocular, borne in or on a stroma
  A. Locules marginal around a sterile center
    1. Locules immersed
      a. Ascostromata with radiating hyphae at the margin ...... TRICHODOTHIS .... (Trichodothella)
      b. Ascostromata setose or tomentose throughout.............. RHIZOGENE
    2. Locules borne more or less free on the stroma
      a. Locules borne more or less free on the margin .... XENOMERIS
      b. Locules free, pendant from an erumpent tomentose stroma . LASIOBOTRYS
  B. Locules in or on a stroma, not marginal
    1. Locules free, clustered on an erumpent hypostroma .... GIBBERA
    2. Locules sunken in a disc-like or pulvinate stroma
      a. Ascostromata erumpent on a hypostroma
        i. Ascospores septate in the middle .... COCCOIDELLA
        ii. Ascospores septate below the middle.................... COCCOIDEA

b. Ascostromata subcuticular or deeper
 i. Ascospores septate above the middle .... ATOPOSPORA
 ii. Ascospores septate at or below the middle
  (a). Spores dark brown .... CROTONE
  (b). Spores light colored .... PLATYCHORA

Venturia: *Venturia* is a large genus, mostly parasitc upon the leaves of higher plants. The ascostromata are ellipsoid to spheric and imbedded in the host tissue or strongly erumpent. Most species have setae about the ostiole. Some mature their ascostromata on the living host parts, whereas others mature their ascospores on the dead plant parts over winter. They have conidial states in the closely related form genera, *Fusicladium, Spilocaea* and *Pollaccia.*

The most important species is *Venturia inaequalis*. It is the cause of apple scab or scurf, which is very destructive to apple trees wherever they are grown. It is widespread throughout the world, and all apple growers must follow a spraying program in order to have marketable fruit. It usually appears first in the early spring, especially in cool damp weather, in the conidial state on the leaves where it forms a black sooty growth of mycelium below the cuticle. Small stromatic cushions are formed from which clumps of short conidiophores arise. Pyriform, one- or two-celled, brown conidia are formed as "blown out" conidia from the apex of the conidiophore. When the conidium falls off it leaves a ring-like scar, through which a second conidium forms. This results in a series of ring-like scars about the apex of the conidiophore. This conidial state has been known in the literature as *Fusicladium dendriticum*. Hughes (**1953**), however has shown that this type of annular conidial formation places this conidial state in the genus *Spilocaea,* as *S. pomi.*

The development of the ascostroma has been reported by Killian (**1917**). The ascostroma originates just beneath the lower epidermis as the coiled tip af a vegetative hypha. This hypha coils about itself to form a hyphal knot. The outer cells of this knot become thick-walled and pigmented and give rise to the wall of the fruit body. A central cell takes a deeper stain and shows larger nuclei. This cell divides and the nuclei increase in number to form a coiled archicarp. The tip of this archicarp elongates and grows out through the wall of the hyphal knot into the leaf tissue as a trichogyne. This trichogyne grows toward and contacts a vegetative cell in the leaf tissue. This cell then enlarges and differentiates into an antheridium. There is then a fusion between the trichogyne and the antheridium and "male" nuclei of the antheridium pass down the trichogyne into the basal cells of the archicarp. Paired nuclei then appear in this ascogenous cell, which becomes lobed and grows out as a system of ascogenous hyphae. Killian described paraphyses as originating from the inner wall cells of the locule. The asci arise from croziers and grow up among and disintegrate these sterile filaments. The ascospores mature on the dead leaves and are distributed in the spring rains. This life cycle determines the spray program which all orchardists must follow. Agricultural Eperiment Stations expose greased slides in the spring to catch airborne spores. The time of spore dissemination is broadcast and orchard trees are sprayed every ten days from this time onward. Secondary infections occur on the fruits if this program is not followed and cause scabby contorted spots on the fruits.

Keitt and his associates (Bachus and Keitt, **1940**; Boone, Kline, and Keitt, **1957**; Keitt and Langford, **1941**; Keitt. Langford, and Shay, **1943**: Leben and Shay, **1948**) have published numerous articles on the genetics of *Venturia inaequalis*. The species was shown to be heterothallic. They were able to culture the fungus on decoctions of dead apple leaves and by crossing the plus and minus strains obtained ascostromata in culture. Their cultures were started at 20 C, after ten days they were placed at 8 C, as a cold period was needed to mature asci. When returned to 15 C mature ascospores were formed. They noticed two types of infection. Some "sporulating" strains produced numerous conidia, but others called "fleck" were without conidia. A fleck strain on one variety of apple might sporulate and form conidia on a second strain of apple, but the strain type was constant on any given variety of apple (Kline et al., **1957**). They were able to produce ascospores in culture and, thus, carry the fungus through the parasitic stage. Various other morphologic and physiologic mutants were obtained. They obtained various mutants deficient in certain vitamins, purines and pyridines as related in *Neurospora* (see that genus). Tests showed that all those strains deficient in riboflavin, nitrogen bases and amino acids were lacking in their ability to infect (Pelletier and Keitt, **1954**). Deficient mutants when supplied with the deficient substance, however, gave pathogenic reactions.

Venturia has bitunicate asci and has been placed in the Pleosporales of the Loculoascomycetes. However, this is another case in which there are certain anomalies. Killian has has shown that the ascocarp arises as a coiled hypha on a vegetative mycelium as in the Ascohymeniales. The wall cells arise from hyphae which grow around this coil. Furthermore, Killian described true paraphyses arising from the wall cells of the locule. These are characters of a true perithecium and not of an ascostroma.

The scab organism of pear has been placed in a separate species, *V. pirina*. Its conidial state is a true *Fusicladium.* In this case the conidia are one-celled, brown and formed on short sterigmata on the apex of the conidiophore. When a conidium falls off, the apex of the conidiophore grows to one side and a second sterigma with a conidium is formed. This results in a conidiophore with a number of peg-like sterigmata on which deciduous conidia have been formed.

Venturia tremulae with *Pollaccia radiata* as its conidial state, and *V. populina* with *P. elegans* as its imperfect state, cause a shoot blight of *Populus*, in Europe and America respectively (Dance, **1959, 1961**; Gremmen, **1951**).

Gibbera: The genus *Gibbera* has been variously interpreted. The ascostromatic characters are extremely variable, even in a single species, and many genera have been described which are closely related. Von Arx and Müller (**1954**), themselves, have interpreted the genus variously, but lately (Müller and von Arx, **1962**) they have limited the genus to species with a more or less well-developed hypostroma and no surface mycelium. The ascostromata are often setose. The ascospores are septate in the middle and greenish to olive-brown.

Seynesiella: This genus has subcuticular, superficially erumpent ascostromata with a flattened base, hemispheric in form and with a marginal radiate upper wall. In these characters the genus resembles the Microthyriaceae.

Antennularia, Acantharia, Pseudoparodia and *Apiosporina* have superficial ascostromata borne on or in a superficial subiculum. In *Atennularia* the ascostromata may be tomentose or setose. In *Acantharia*, setae occur on the ascostromata and also on the subiculum. In *Pseudoparodia* the ascostromata open widely at maturity in a discomycete-like fashion.

Apiosporina: Apiosporina collinsii (Sprague and Heald, **1927**) is a parasite on species on *Amelanchier*. It forms a perennial, systemic mycelium which causes a witches broom. It is widely distributed throughout North America wherever species of *Amelanchier* occur. In early spring, leaves on the branches of the witches broom show an olive colored growth which consists of small clumps of elongate conidiophores bearing chains of one-celled, brown, ellipsoid conidia of the form genus *Cladosporium*. As the season advances, a dark sooty growth of mycelium occurs on the under side of the leaves. The ascostromata arise in large numbers, forming a crust on this subiculum.

The development of the ascostromata has been described by Sartoris and Kauffman (**1925**). The ascostroma arises about the first of July as a knot of intertwined hyphae. The central cells of this knot take a deep stain and are uninucleate. The outer hyphae become compacted to form the wall of the fruit body. The inner cells divide and eventually send out a fan-shaped group of hyphae which reach the top of the locule. About this time certain basal cells show a terminal binucleate cell. The two nuclei of this terminal cell undergo a conjugate division. Two septa are laid down in such a way that an apical, uninucleate cell, a central binucleate cell and a basal uninucleate cell are formed, much as in crozier formation. The apical cell continues to grow out as elongate filaments which soon degenerate. The central binucleate cell becomes the young ascus. Asci are formed in September, but ascospores do not mature until the following spring. The ascospores are delimited in the two-nucleate stage, the septum being asymmetrically placed cutting off a smaller lower cell and a larger upper cell.

In late fall pyriform pycnidia appear on the surface of the infected leaves. They contain filiform conidiophores which cut off fusoid conidia.

In *Apiosporina* the ascostromata have short peg-like setae. In *Metacoleroa* the ascostromata are seated on a sheet-like mycelium and the setae are long and spine-like.

Phaeocryptopus: Phaeocryptopus nudus is a parasite of needles of fir. It forms tiny superficial ascostromata each with a short stalk arising from a subepidermal hypostroma. *Phaeocryptopus pinastri* occurs on pine needles.

Botryostroma: Botryostroma is similar in habit to *Parodiella*. The ascostroma is superficial with a flattened base and the ascospores are apiosporous with a papillate lower cell, and hyaline to light brown.

The genera *Xenomeris, Trichodothis, Trichodothella, Rhizogene, Lasiobotrys, Coccoidea* and *Coccoidella* all have superficial ascostromata arising from a hypostroma within the host. The stromata are coarsely parenchymatic and these genera have been placed in the Dothideaceae in this account.

Xenomeris: This genus is similar to *Gibbera*, but the superficial ascostromata are more flattened and the locules are in superficial outgrowths which are marginal in their distribution. *Trichodothis* and *Trichodothella* have a central sterile foot and column, around which the locules are oriented. The stromata are flattened and have tomentose hyphae radiating from the margin. In *Trichodothis* the locules are imbedded around the sterile center. In *Trichodothella* the locules are freer and upon the margin of the stroma.

Rhizogene: This genus has a crust-like superficial ascostroma with sunken locules in the margin and tomentose surface.

Lasiobotrys: In *Lasiobotrys* there is a disc-like superficial ascostroma with tomentum, but the locules are in separate spherical ascostromata which are pendant on the underside of the stroma.

Coccoidea and *Coccoidella* have superficial disc-like or crustose ascostromata with a central hypostromatic foot. The locules are sunken in the stroma, which is not tomentose. *Coccoidella* has ascospores with a central septum whereas those of *Coccoidea* are apiosporous with a very small lower cell.

Crotone, Atopospora and *Platychora* have immersed or erumpent ascostromata. In *Crotone* the ascostromata are erumpent on hypertrophic swellings on the host leaves and the spores are septate below the middle. In *Platychora* the ascostromata are subepidermal, crust-like and the ascospores are also septate below the middle. In *Atopospora* the crust-like ascostromata are subcuticular and the ascospores are septate above the middle.

DIMERIACEAE:

The family Dimeriaceae is recognized by Müller and von Arx (**1962**). It consists of forms having unilocular ascostromata which are superficial parasites on lichens, mosses, and the leaves of higher plants as well as hyperparasites on other fungi such as the Meliolaceae. The genera included here were originally placed in the Perisporiales, because of their obscure ostioles. These genera are similar in their ecological habit to the Erysiphaceae, Meliolaceae and Perisporiopsidaceae. They differ from the latter in their more cylindric and more numerous asci and in their papillate rather than perforate ostioles. Hansford (**1946**) placed all of these forms in the Sphaeriaceae. The ascostromata may be glabrous, setose or tomentose. The ascospores are two-celled, yellow, greenish or brown, in which characters they approach such genera as *Antennularia* and *Apiosporina* in the Venturiaceae. Many species have a black sooty superficial growth which places them in the general group of the sooty molds. Certain genera of the Capnodiaceae (*Limacinia, Echinothecium*) differ chiefly in the torulose mycelium. They are mostly southern in their distribution.

I. Ascostromata superficial, stalked .... PODODIMERIA
II. Ascostromata superficial, not stalked
  A. Ascostroma large (180-300 μ); ascospores brown, on moss .... LIZONIA
  B. Leaf parasites, ascostromata 200 μ or less; with intramatrical mycelium
    1. Superficial subiculum with tomentose hyphae on mycelium and ascostromata .... EUMELA
    2. Superficial mycelium scanty; ascostromata not tomentose... EPISPHAERELLA
  C. Superficial epiphytes or hyperparasites; mycelium intracuticular not intramatrical
    1. Hyperparasites on fungus mycelium, wall of a single layer of cells. Phomoid conidia present
      a. Ascostromata with setae .... PHAEODIMERIELLA
      b. Ascostromata glabrous
        i. Ascospores remaining hyaline......................... DIMERINA
        ii. Ascospores becoming brown .... DIMERIUM

2. Epiphytes, ascostromata upon their own mycelium; wall of several layers of cells
   a. Ascospores becoming greenish-yellow .... EPIPOLAEUM
   b. Ascospores remaining hyaline
      i. Ascostromata with lobed setae .... WENTIOMYCES
      ii. Ascostromata glabrous, with a subiculum.......... EUDIMERIOLUM

Dimerium: The genus *Dimerium* is a hyperparasite on the mycelium of species of the Meliolaceae and the Asterinaceae. The ascospores are hyaline but become brown at maturity. In some species (*Bolosphaera*) the ascocarps are bluish or reddish. The form genera *Cicinnobella* and *Ectosticta* are given as conidial states.

There is a group of dimeriaceous fungi which occur on the needles of conifers, causing a needle blight. They were all placed in the same species, *Dimeriella balsamicola* by Farr (**1963**). Shoemaker (**1964**) examined the type collection of this species and found that it grew on the mycelium of a *Meliola*, and cylindric asci and small spores (7-9 × 5.5 μ) and glabrous ascostromata. He retained four closely related species for the other fungi concerned. These species had broad clavate asci with a thickened apical wall and larger spores (9-36 × 2.5-5 μ). They were placed in the genus *Epipolaeum*. The four species differ in the size of the spores and the ascostromata. They occur on *Abies, Tsuga,* and *Pseudotsuga. Phaeocryptopus* and an asterinaceous fungus were also associated with the blights on *Abies*. The ascostromata and ascospores of these fungi mature slowly and need a moist atmosphere for growth. As a result, the ascospores are usually immature and hyaline, but become brown when fully mature or near germination.

Phaeodimeriella: This genus differs from *Dimerium* chiefly in the setose ascostromata. It also has pycnidia with elongate conidia which are hyaline in *Ectosticta* and brownish in *Cicinnobella.*

Dimerina: This differs from the two preceding genera in its hyaline spores. It also has *Ectosticta* pycnidia as a conidial state.

Epipolaeum: This has a superficial mycelium and glabrous tomentose or setose ascostromata. Some setae are slightly lobed as in *Wentiomyces*. The ascospores become greenish or brownish at maturity.

Eumela and *Episphaerella*: These genera have both a superficial mycelium and appresoria on hyphopodia which penetrate into the epidermis or deeper. *Eumela* has tomentose ascostromata; *Episphaerella* has not.

Pododimeria: This genus has glabrous ascostromata upon a definite stalk.

LOPHIOSTOMATACEAE:

The Lophiostomataceae include forms with immersed to superficial unilocular ascostromata which have a flattened ostiolar neck which opens by a longitudinal slit. They were usually considered as being intermediate between the Sphaeriaceae and the Hysteriales. Practically nothing is known as to the details of the development of the ascostroma and the interthecial tissues. Munk (**1957**) placed them in his Pleosporaceae. Müller and von Arx (**1950**) considered *Lophiostoma* as closely related to *Leptosphaeria*. Later (**1962**) they said the asci of the Lophiostomaceae were bitunicate and placed the family in their Pseudosphaeriales. The placement, here, in the Pleosporales is arbitrary, for more detailed information as to their development is needed. The species usually occur on woody substrata as saprophytes.

I. Ascostromata on a subiculum.................................BYSSOLOPHIS
II. Ascostromata not in a subiculum
  A. Spores 1-celled, brown...LOPHIELLA
  B. Spores 2-celled
    1. Spores appendaged at each end .... LAMBOTTIELLA
    2. Spores not appendaged
      a. Spores hyaline .................................LOPHIOSPHAERA
      b. Spores brown.......................................OSTROPELLA
      ....(Schizostoma)
  C. Spores 2- to many-celled
    1. Spores hyaline
      a. Spores appendaged at each end.........................VIVIANELLA
      b. Spores not appendaged.............................LOPHIOTREMA
    2. Spores brown
      a. Spores appendaged at each end.......................BRIGANTIELLA
      b. Spores not appendaged.............................LOPHIOSTOMA
  D. Spores muriform
    1. Spores hyaline
      a. Spores caudate at base.................................SAMPAIOA
      b. Spores not appendaged ...........................LOPHIDIOPSIS
    2. Spores brown ....PLATYSTOMUM
  E. Spores filiform ....LOPHIONEMA

MESNIERACEAE:

This is a small family of leaf parasites based on the brown ascospores which have sculptured outer walls consisting of longitudinal striae (except in *Butleria*). The ascostromata are unilocular, immersed, light colored to yellow and with small papillate ostioles. Von Arx and Müller (**1954**) placed them in the Dothiorales, but later (Müller and von Arx, **1962**) transferred them to their Pseudosphaeriales. The asci are bitunicate and contain four to sixteen ascospores. There are numerous paraphysoids (?).

I. Spores 1-celled .... MESNIERA
II. Spores 2-celled
  A. Ascospores with a sculptured wall .... STEGASPHAERIA
  B. Ascospore wall not sculptured .... BUTLERIA

CORYNELIACEAE:

The Family Coryneliaceae is a curious one and of doubtful relationship. The ascostromata occur in superficial clusters arising from erumpent stromata, mostly on the leaves of higher plants. The ascostromata are stalked and have a more or less elongated ostiolar neck. A single locule occurs in each stalked ascostroma (except in *Sorica*, which may have several). The ostiolar neck opens either by a transverse or radiate slit or by a perforation. The asci are broad clavate, with a long filiform stalk, and lie at various levels in the locule. The ascospores are freed within the locule and ejected as a sphaerical mass of spores at the mouth of the ostiole. Most of the species are found in the southern hemisphere. *Corynelia, Tripospora* and *Lagenulopsis* occur on the leaves of *Podocarpus*. *Sorica* occurs on the leaves of the fern *Polypodium*. *Caliciopsis* is found on the needles and bark of conifers, and *Acrospermum* on the leaves of grasses and dicotyledons.

Fitzpatrick (**1942**) considers the ascocarp of the Coryneliaceae to be stromatic. There is no perithecial wall and the asci arise in a locule in a stroma. No paraphyses are present. On the other hand the asci, although without an apical apparatus, are thin-walled and do not appear to be bitunicate. The family has been considered by Arnaud (**1931**) and others to be related to the Discomycetes. However, the asci are borne in a fascicle in the base of the locule and the ascostromata open by slit or pore. These genera have also been placed in the Hysteriales, but the ascostromata are vertically upright and not linear and most genera have a perforate ostiole. Arnaud (**1931**) erected a family Caliciacees in which he included the following genera in addition to those in the Coryneliaceae:

A. Apothecia stalked, with the asci over a convex surface of a spherical head (Coniocybée) .... CONOCYBE
B. Apothecia stalked, more or less enclosed, on lichens. Ascospores 1-celled, brown (Calicées)
  1. Apothecia with a short stalk .... SPHINCTRINA

2. Apothecia long stalked
   a. Ascospores globose .... CYPHELIUM
   b. Ascospores ellipsoid .... EMBOLUS

I. Ascospores 2-celled, brown
   1. Apothecia pedicellate .... CALICIUM
   2. Apothecia sessile .... ACOLIUM

II. Ascospores filiform .... ACROSPERMUM

The above genera have a more wide open apothecium-like ascocarp. Arnaud included in his Corynelièes, which correspond to the Coryneliaceae of Fitzpatrick, the following genera:

I. Ascocarps borne singly, spores filiform .... ACROSPERMUM

II. Ascostromata borne in clumps, spores not filiform
   A. Ascostroma with one locule
      1. Ostiolar neck lobed by one or more deep slits; ascospores thick-walled, globose, echinulate, with germ pores .... CORYNELIA
      2. Apex of ostiolar neck perforate, funnel-form
         a. Spores with 4 stout radiating lobes .... TRIPOSPORA
         b. Spores globose to ellipsoid
            i. Spores thick-walled, echinulate, 10.5 μ in diameter... CORYNELIOPSIS
            ii. Spores smooth, thick-walled
               (a). Asci 2-spored; spores 11-15 μ in diameter on *Podocarpus* .... LAGENULOPSIS
               (b). Asci 8-spored, ascospores thin-walled, 4-10 × 2.5 μ; not on *Podocarpus* .... CALICIOPSIS
   B. Ascostroma proliferating; forming more than 1 locule .... SORICA

III. Ascostromata and spermagonia on a coarse subiculum of rope-like hyphae .. SCORIAS

The family has been monographed by Fitzpatrick (**1942, 1942a**).

Corynelia: This forms clumps of short stalked ascostromata with short stout ostiolar necks which open by one or more deep slits. The ascospores are spherical, dark brown, echinulate and have prominent germ pores. It occurs on *Podocarpus* leaves.

Tripospora: This has similar clumped ascostromata on leaves of *Podocarpus*. The ostiolar necks are longer and open by a pore forming a funnel, which becomes filled with a mass of pulvinate ascospores at maturity. The ascospores are hyaline to dark brown, thick-walled and have a central globular portion with four-five radiating lobes.

Lagenulopsis: This has ascostromata similar to *Tripospora*, with filiform ostiolar necks. The ascospores are globose with a thick, smooth dark brown wall.

Coryneliospora: This genus has ascospores as in *Corynelia*, but spheric to ovoid, and the ostiolar necks open by a pore rather than by a slit. The species occur on plants other than *Podocarpus*.

Sorica: This is parasitic on the fern, *Polypodium*. It has dense clumps of ascostroma with much elongated stalks and ostiolar neck. The ascocarp shows proliferation, that is, more than one locule may be formed in the long filiform stalk. Fitzpatrick included the genus in *Caliciopsis*.

Caliciopsis: This genus occurs on the leaves and branches of conifers and some dicotyledons. The ascostromata are clustered with long filiform ostiolar necks. There is a slight swelling where the ascus locule occurs. The ascospores are smaller and thinner walled than in *Corynelia*. *Caliciopsis pinea*, which is found on cankers of white pine, was cultured by Ray (**1936**), who obtained spermagonia and a few ascocarps in culture.

McCormack (**1936**) studied the formation of the fruit bodies of *C. pinea*. Confluent stromata originate beneath the surface of the bark. These stromata become erumpent and give rise to lobes which are the initials of spermagonia and ascostromata. The spermagonia arise as cavities in these lobes and are formed by the disintegration of the locule tissues. Rod-shaped to allantoid spermatia are cut off from bottle-shaped conidiophores. Ascocarps arise in similar lobes in cavities filled with hyaline hyphae. One or more coiled, deeply staining archicarps were seen in such cavities. The cells of these archicarps divide and enlarge. As the ascostromatic lobe elongates, the archicarps disappear and the center of the cavity becomes filled with ascogenous hyphae. The ascogenous hyphae branch profusely and the cavity is extended upward above them. The asci from the ascogenous hyphae are clavate and have long filiform stalks. These form a fascicle in the base of the locule. The asci deliquesce and free the ascospores in the locule centrum, where they are pushed out through the ostiolar neck as a plug of globose golden brown ascospores.

Acrospermum: In *Acrospermum* the ascostromata occur singly or in small groups, superficially on the leaves of various higher plants. The asci are long clavate-filiform as are the spores, which are nearly as long as asci. The development of the ascostroma in *A. compressum* was studied by Brandriff (**1936**). The ascostroma arises as a homogenous parenchymatic, stromatic lobe. As this lobe elongates a cavity arises within it by the loosening of the hyphae. No archicarp was observed. The base of the cavity becomes filled with ascogenous hyphae. Paraphyses are formed which grow upward. The asci grow up through the paraphyses. A pore-like ostiole is formed at the apex of the lobe. The development is similar to that reported for *Caliciopsis*.

Webster (**1956**) considered *A. compressum*, which occurs on leaves of dicotyledonous plants, and *A. graminum*, which occurs on grasses, as distinct species on the basis of their conidial states produced in culture. In *A. compressum*, upright conidiophores were formed. Two-celled, cylindric, hyaline conidia were cut off successively from the apex of such conidiophores, forming a cluster of conidia at the apex. In *A. graminum*, short conidiophores were produced from rope-like strands of hyphae. The conidia in this species were shorter, one-celled, cylindric and hyaline and formed in similar clusters at the apex of the conidiophores.

SPORORMIACEAE:

The genera of this family were formerly placed in the Sordariaceae, because of their occurrence on dung. Cain (**1934**) found that the ascus in these genera have a double wall and eject their ascospores in the manner of a bitunicate ascus. Consequently, Müller and von Arx (**1962**) placed these genera in a family Sporormiaceae, in their Pseudosphaeriales. The ascostromata are unilocular, membranous and more or less immersed in the substratum. The ascospores are deep brown, two- to many-celled, and have a vertical or diagonal germ slit in each cell, or a germ pore (Arx NS Müller, **1955**).

I. Ascospores 2-celled
  A. Ascostromata glabrous; ascospores with germ slits .... DELITSCHIA
  B. Ascostromata setose about the ostiole; ascospores with a germ pore at each end .... TRICHODELITSCHIA
II. Ascospores with more than 2 cells; with germ slits ....SPORORMIA
III. Ascospores muriform; with germ slits .... PLEOPHRAGMIA

Delitschia: The ascostromata of this genus are sunken in the substrate. The Ascospores are two-celled, constricted and have a diagonal germ slit and a gelatinous envelope. The cells may fall apart in some species. Species with sixteen ascospores in an ascus have been placed in a genus *Delitschiella* and those with two ascospores in *Pachyspora*, but the number of ascospores is variable.

Trichodelitschia: In this genus there is a crown of setae about the ostiolar neck and the two-celled ascospores have a germ pore at each end.

Sporormia: In this genus the spores are three- to many-septate and each cell has a diagonal or vertical germ slit (Arnold, **1928**). The genus *Preussia*, which is placed in the Plectascales has spores very similar to *Sporormia* and probably represents a nonostiolate form of this genus. The spores usually have a gelatinous envelope.

Pleophragmia: The ascospores in this genus are muriform with vertical as well as transverse septa. Von Höhnel placed the species in *Pleospora*, but the spores are quite different from those of that genus and have a gelatinous envelope.

MASSARIACEAE:

The family Massariaceae was originally erected for species having spores with gelatinous envelopes. This is an entirely arbitrary character for gelatinous envelopes occur on spores of species having different affinities. However, the type species of *Massaria, M. inquinans*, is a distinct form which can stand as the type of this group. These *Massaria* species have rather large ascostromata with leathery walls of small-celled parenchyma, immersed in the host tissue. The ascostromata have stout, cylindric, separately erumpent ostioles and are surrounded by blackened zones which extend from one ascostroma to the next within the host tissue. In the genus *Massarina*, there may be some clypeate development of hyphae in the surface host tissues, which is all that remains of this stromatic development. The spores of the *M. inquinans* group are characteristic. They are elongate-fusoid and surrounded by gelatinous envelopes. They are four-celled, with the central cells flattened and the two end cells longer and tapered at the ends. Each cell contains a large conspicuous oil drop. The paraphysoids in the ascostromata are fine filiform and very numerous. The ascospores are usually brown but may be hyaline in some species.

The genus *Massarina* was originally described as a *Massaria* with hyaline spores. Bose (**1961**) has monographed the genus. The ascostromata are similar to those of *Massaria*, but they lack the outlining blackened zones found in *Massaria*. They are immersed, have walls of small-celled parenchyma, and erumpent ostioles which may have some adjacent clypeate development. The paraphysoids are numerous and filiform. Bose showed these filaments attached at both top and bottom and arising before the formation of asci. He saw the ascogenous hyphae arising from the base of the paraphysoids, and the asci growing up between them. The paraphysoids become free at their apices at maturity. I have seen *Phoma*-like pycnidia in cultures of *Massaria.* Bose reports pycnidia of the form genera *Ceratophoma, Dendrophoma, Microsphaeropsis, Coniothyrium* and *Diplodia* as imperfect states of *Massarina*. The ascospores of *Massarina* are usually one-septate at first, but become three-septate at maturity. They are long fusoid, constricted at the septa, hyaline and sometimes with gelatinous envelopes. Bose includes the genus *Massariella*, which was described as having two-celled ascospores. These species of *Massarina* could well be the origin of the *Massaria inquinans* type.

I. Spores 4-celled, central cells flat, end cells tapered, hyaline or brown. Ascostromata outlined by a blackened zone. . . . . MASSARIA

II. Spores 2-4-celled, hyaline, cells approximately equal. Stromata with clypeus but no blackened zone . . . . MASSARINA

MICROTHYRIALES

(Hemisphaeriales)

This order was originally called the Hemisphaeriales (Theissen, **1913**; Theissen and Sydow, **1915**). Arnaud (**1925**) used the name Microthyriales for these dimidiate forms and this has been followed by Luttrell (**1955**) and others.

The characteristics which unite the many taxa in this order are found in the ascocarps, often called "thyrothecia," which are shield-shaped (dimidiate), elongate or sometimes irregular and confluent. Most of them have a radiate upper wall or shield (except in the Hemisphaeriaceae). The lower wall is absent or poorly developed, and its protective function is supplied by the epidermis or other surface tissue of the host or substrate.

A large number of families and tribes has been described by various authors. Many of the genera are poorly known and their transfer from one family to another has caused a great deal of confusion. The following is a rough attempt to key out the many families and tribes which have been described.

I. Upper wall (shield) radiate
  A. Ascostromata borne under a superficial, plate-like thallus
    1. Thallus composed of parallel hyphae................. TRICHOPELTACEAE
    2. Thallus composed of radiating cells..................... BREFELDIACEAE
  B. Ascostromata not borne beneath such a thallus
    1. Upper tissues of ascostromata gelatinizing at maturity .... ENGLERULACEAE
    2. Upper tissues of ascostromata not gelatinizing at maturity
      a. Interthecial tissue parenchymatic............. STEPHANOTHECACEAE
      .... (Microthyriellaceae)
      b. Interthecial tissue other than parenchymatic
        i. Ascostroma superficial, with a surface mycelium
          (a). Opening by an ostiole................... MICROTHYRIACEAE
          (b). Opening by a slit........................... ASTERINACEAE
        ii. Ascostroma superficial, but anchored by a hypostroma.............
          .... POLYSTOMELLACEAE
        iii. Ascostromata subcuticular
          (a). On Dicotyledons........................ STIGMATEACEAE
          (b). On ferns and conifers...................... LEPTOPELTACEAE
        iv. Ascostromata borne on a mycelial hypha, both upper and lower shield radiate; hyperparasites .... TRICHOTHYRIACEAE

II. Upper wall not radiate
  A. Ascospores 1-celled, brown, with equatorial band............ ENTOPELTACEAE
  B. Ascospres various, not as above....................... HEMISPHAERIACEAE
  .... (Micropeltaceae)

The members of this order are superficial parasites, hyperparasites or saprophytes upon the leaves and stems of higher plants. There is very little internal mycelium, often limited to haustoria in the epidermal cells.

The superficial mycelium, when present, is usually dark brown, giving a sooty appearance to the infected spots. The hyphae may contain somewhat swollen cells known as node cells (stigmatocysts) or short modified branches known as "hyphopodia," from which the haustoria arise. There may or may not be superficial mycelium, which may or may not have hyphopodia or node cells. Setae may occur on the mycelia or ascocarps of some genera.

Although the order Microthyriales appears as a unit because of the thyrothecial type of ascocarp, there is a great deal of variation in many secondary characters which have been used to separate genera.

Various types of conidia on the mycelia or in pycnidia are known.

The ascocarps themselves may be shield-shaped (dimidiate), elongate, irregular in shape or confluent. The upper wall or "shield" may be "radiate," that is, consisting of dichotomously branched hyphae radiating from a central point; "meandrous," consisting of meandering, irregularly radiating hyphae; or "plectenchymatous," consisting of irregular shaped cells, *textura epidermoidea*, somewhat in the pattern of the epidermis of some higher plants. The lower wall is usually absent or poorly developed (except in the Trichopeltaceae) and its protective function is supplied by the epidermis of the host. The upper wall may open by a perforate ostiole, a radiate or elongate split, or by the gelatinization of the upper tissue.

The ascus is bitunicate, saccate to clavate and with a thickened apical wall. The ascospores are various but mostly two-celled and often brown.

There have been very few developmental studies in the Microthyriaceae. Ryan (**1926**) noted that most thyrothecia arise from a single intercalary cell, a short side branch or a hyphopodium. This cell divides into a few isodiametric cells which send out radiating hyphae which fuse to form the upper shield. Luttrell (**1940, 1944a, 1946**) has given a more detailed study of ascocarp development in several species which will be discussed below. He found that some species of the Microthyriales have asci formed in individual locules as in his *Pseudosphaeria* type. He erected the family Microthyriellaceae for such (myriangiaceous) species, but later accepted Petrak's (**1931**) Stephanothecaceae as a synonym. He placed all radiate forms in the family Microthyriaceae and all nonradiate forms in the family Micropeltaceae (Hemisphaeriaceae). Luttrell (**1946**) suggested that all genera described as having paraphyses should be placed in his Microthyriaceae and those described as lacking paraphyses in his Stephanothecaceae.

There have been several score or more genera described in the Microthyriales, many of which have been transferred back and forth between families and orders.

Keys to the genera of the Microthyriales may be found in Theissen (**1916**), Theissen and Sydow (**1915**), Stevens (**1918**), Doidge (**1942**), von Arx and Müller (**1954**), and Müller and von Arx (**1962**).

Very little is known concerning the origin of the interthecial tissue in the Microthyriales. Many genera have species both with and without paraphysoids or pseudoparaphyses.

Conidial states are usually found in pycnidia which resemble the thyrothecia in being flat and radiate. The form genera belong mostly in the Pycnothyrieae of the Leptostromataceae. The genus *Asterina* has conidial states in the form genus *Asterostomella*. The conidia in the pycnidia are one-celled, brown, and often with a light colored equatorial band. The form genera *Asteromella* and *Actinothyrium* also occur. *Echidnodes* has pycnidia with ovoid to fusoid conidia which are one- to three-septate. *Clypeolella* has three-septate brown conidia borne on lateral branches of the mycelium. Spermagonia (pycnidia with bacillus-like conidia) are also reported from a number of genera (*Myiocopron, Lembosia, Lembosiopsis*).

In such genera as *Englerula* and *Englerulaster* the thyrothecia open by a gelatinization of the upper tissues, as in the Englerulaceae (Petrak, **1928a**).

STEPHANOTHECACEAE:

This is the family in which Luttrell would place all of the thyrothecial forms having asci imbedded singly in the stromatic tissue. He would include all genera described as lacking paraphyses.

Luttrell (**1940**) described ascocarp development in *Morenoella quercina*, which causes a black spot disease on leaves of oak. It has superficial dark brown mycelium upon which circular to elongate sometimes confluent spots arise in early summer. Spermagonia appear in August and are mature in September. They arise from hyphopodia and form a hemispherical shield which is radiate at first. Beneath this shield a stromatic tissue arises without any basal wall. Small rod-like spermatia are cut off from this tissue. Ascocarps arise in the same manner from hyphopodia. Hyphae grow out from the under side of the shield to form a loose plectenchyma, in which ascogenous hyphae arise. The asci, as they are formed, crowd out the intervening tissue and lie in a bed of parenchyma. No fertilizing function was detected for the spermatia.

Schizothyrium, which is placed in the Schizothyriaceae by Müller and von Arx (**1962**), has asci imbedded singly in a stromatic matrix, and belongs in this family.

MICROTHYRIACEAE:

Müller and von Arx (**1962**) include in this family those species with a radiate shield and an ostiolar opening. Luttrell (**1955**) would include all species with a radiate shield. Luttrell (**1944**) has described the life history of *Myiocopron smilacis*. The fungus infects living stems of *Smilax* but fruits only on the dead portions of the stems. Spermagonial and ascocarp initials arise about one month after the stem dies. The ascocarps mature about six months later. Spermagonia arise from a superficial hypha or from a hypha which grows up from a spherical mass of mycelium formed within the host. A plate of spherical cells is formed from this superficial cell and branches from these cells grow out radially with a dichotomous branching to form a radiate shield. Fine hyphae grow downward from the underside of the shield to form a parenchymatous stroma. Spermatiophores arise from the roof of this shield and grow downward and form two types of hyaline spermatia, one allantoid and the other more elongate cylindric.

The ascostromata arise in the same manner as the spermagonia. Hyphae from the underside of the shield form a thin parenchymatic layer. An ascogonium, consisting of a coiled deep staining hypha, arises in this parenchyma. The ascogonium proliferates to form a mass of swollen ascogenous hyphae. Parenchyma cells above the ascogonium become meristematic and produce filiform hyphae which grow downward. The tips of these hyphae grow around the base of the ascogenous hyphae forming a loose prosenchyma. The asci arise without croziers and grow up between these filiform hyphae which are of the nature of pseudoparaphyses. These pseudoparaphyses differ from the paraphysoids supposedly characteristic of the Pleosporales, in their downward growth from above. This is the sort of interthecial tissue found in the Hypocreales of the Unitunicatae.

ASTERINACEAE:

Various genera have been placed in this family. Müller and von Arx limit it to radiate forms with a superficial mycelium and a stellate opening of the thyrothecium. It would fall in the Microthyriaceae of Luttrell. Some species of *Asterina*, opening by the gelatinization of the upper tissues, resemble the Englerulaceae.

POLYSTOMELLACEAE:

This family is based upon the presence of a foot or hypostroma which anchors the thyrothecium within the host. It is the Parmulariaceae of Müller and von Arx (**1962**). Many of the species of this family have plurilocular stromata. The locules may be elongate and open by a slit, or circular with an irregular opening. The locules may be arranged radially, in concentric circles about a central sterile plug, or scattered irregularly. Luttrell and others do not consider the presence of a hypostroma to warrant family rank. In fact ascostromata with or without such a foot may be found even in the same species (as in *Myiocopron*).

HEMISPHAERIACEAE (Micropeltaceae):

In this family the shield is not radiate but the hyphae form a meandering plectenchymatic pattern. *Stomiopeltis polyloculatis,* which causes an olive blotch of *Arundinaria* canes, was studied by Luttrell. It has a dark brown surface mycelium. Young ascocarps arise during July. Asci appear in October and remain in the fusion nucleus stage in winter. They mature the following April through June. The ascocarps arise by the twisting, branching and anastomosing of the hyphae of the superficial mycelium at various points. This forms a nonradiate upper shield. Luttrell would limit *Stomiopeltis* to such nonradiate species. Two or three layers of thin-walled parenchyma cells are formed in the underside of this shield. At intervals over this shield some cells become meristematic and give rise to a number of ascus locules. These meristematic cells give rise to filiform hyphae which grow downward and bend inward along the cuticle to form the lower wall of the thyrothecium. During September, binucleate cells appear at the base of each locule among these "pseudoparaphyses" and grow upward between them. From two to sixteen such locules may be formed beneath one shield. The ascospores are two-celled and hyaline. No spermagonia are formed.

STIGMATEACEAE:

In this family there are placed those species in which the ascostroma originates beneath the cuticle. Killian (**1922**) followed the development of *Stigmatea robertiana* on leaves of *Geranium*. The fungus forms a layer of mycelium beneath the cuticle on the surface of the epidermis. This mycelium thickens at various points to form a pseudoparenchyma within this stroma, two large deeply staining cells with large nuclei appear, and represent the archicarps. The nuclei of these cells divide followed by cell division to form a group of cells. One of the central cells produces an elongation which Killian believes to fuse with a neighboring antheridial hypha. The number of nuclei increases within the ascogenous cell. These nuclei migrate in pairs into ascogenous hyphae which grow out from the ascogenous cells. Asci are formed from croziers. The interthecial tissues of the stroma elongate to form filaments. The exact method of formation of these filaments is not clear from Killian's figures. This indicates the difficulty of interpreting the method of formation of such filaments without exact information, for there are variations of the so-called types.

TRICHOPELTACEAE:

In this family the only superficial mycelium is a thallus consisting of a radial plate or branching ribbons of cells upon which the ascostromata are formed. Species with a linear or branched thallus are placed in the Brefeldineae and those with a circular thallus in the Trichopeltineae.

TRICHOTHYRIACEAE:

The species in this family are hyperparasites on other fungi. The thyrothecia arise from an aerial hypha. The radial upper shield is formed as in the other species of the Microthyriales, and a lower radial shield is also formed on the under side. The thyrothecia then hang somewhat free from the parent hypha. This has been called an inverted thyrothecium because the ostiole occurs on the upper side beneath the point of attachment to the parent hypha.

LEPTOPELTACEAE:

This family is rather vaguely outlined as being subcuticular or superficial and on ferns or conifers (Arx, **1962**). A generic key is given by Müller and von Arx (**1962**):

I. Spores muriform .... DOTHIOPELTIS
II. Spores 1- to many-celled
    A. Subcuticular, nonradiate, spores 2-celled
        1. Spores hyaline .... LEPTOPELTIS
        2. Spores brown .... THYRIOPSIS
    B. Subcuticular or superficial, radiate
        1. Spores ellipsoid, 1-celled .... MOESZOPELTIS
        2. Spores elongate, fusoid, 2- to many-celled
            a. Spores 2-celled .... LEPTOPELTOPSIS
            b. Spores 3- to 4-celled .... PYCNOTHYRIUM

ENTOPELTACEAE:

This is a small family erected by von Arx and Müller (**1954**) on the basis of the one-celled brown ascospores with a light colored equatorial band. The ascostromata are flat, dimidiate and subcuticular. The upper shield is not radiate. Only three genera are given:

I. Mycelium with an intracuticular, band-like mycelium .... ENTOPELTIS
II. Mycelium without such intracuticular mycelium
  A. Loculi circular, with a central pore .... VIZELLA
  B. Loculi linear opening by a longitudinal slit .... BLASDALEA

From the above discussion, one can see that the Microthyriales are a heterogenous group as to almost every character except the superficial, flat to hemispheric, ascocarps with mostly a radiate upper shield and a poorly developed lower wall. One can see why various genera have been placed in the Dothideales, Pseudosphaeriales, Hysteriales, Sphaeriales and Phacidiales. It seems more logical to place them together in a morphologic and ecologic related group until much more evidence as to their true relationships is obtained.

HYSTERIALES

The genera of the Ascomycetes having ascocarps which open by the formation of an elongate slit or a radiate rupture have been a troublesome group showing relationships to both the Discomycetes and the Pyrenomycetes. The various genera have been transferred back and forth between such families as the Phacidiaceae, Ostropaceae, Tryblidiaceae, Lophodermataceae. Hvsteriaceae and Lophiostomataceae. The two chief families usually placed in the Hysteriales are the Lophodermataceae and the Hysteriaceae. The Lophodermataceae have sunken ascocarps with membranous walls and are obviously related to the Discomycetes, whereas the Hysteriales have superficial fruit bodies with carbonaceous walls and show a relationship to the Lophiostomataceae and the Dothideales. The Hysteriales occur as saprophytes on dry substrata such as weathered wood, bark, resin, etc. The ascocarps are superficial on the substrate and in most genera are ellipsoid to elongate, linear or branched, with a coriaceous wall and open with an elongate slit. In a few genera the ascocarp is conchate, in the shape of a mussel shell set vertically upon edge, with a slit-like opening along the upper edge. Zogg (**1962**) placed these chonchate genera in a separate family, the Lophiaceae and gave a monograph of these two families. The asci are bitunicate, cylindric to clavate and the ascospores are various in color and septation. The conidial states are of the two general types. Pycnidial states with small globose to ellipsoid or fusoid spermatia are found in most genera. Hilitzer (**1929**), who monographed the order, placed all these pycnidial states in a form genus *Hysteropycnis*. Lohman (**1931-1938**), who has published on both the ascus and conidial states of these fungi, placed these pycnidia in various form genera. The second type of conidial state consists of septate colored conidia borne singly or in chains on clumped conidiophores, of the form genera *Septonema, Sirodesmium,* and *Papulospora.*

The following is a key to the genera of this order:

I. Ascospores 1-celled, brown .... FARLOWIELLA
II. Ascospores 2-celled
- A. Ascospores hyaline, ascocarps elongate
  - 1. Ascostromata on a subiculum .... GLONIUM
  - 2. Ascostromata not on a subiculum .... PSILOGLONIUM
- B. Ascospores brown, ascocarps conchate .... BULLIARDELLA

III. Ascospores more than 2-celled
- A. Ascospores hyaline; ascostromata ellipsoid to linear .... GLONIELLA
- B. Ascospores brown
  - 1. Ascostromata ellipsoid to linear .... HYSTERIUM
  - 2. Ascostromata conchate, erect
    - a. Ascospores ellipsoid to cylindric .... MYTILIDION
    - b. Ascospores fusoid .... OSTREION

IV. Ascospores muriform
- A. Ascospores hyaline; ascostromata ellipsoid to elongate .... GLONIOPSIS
- B. Ascospores brown
  - 1. Ascostromata ellipsoid to linear
    - a. With keeled lips along the slit .... HYSTEROCARINA
    - b. Without such a keeled edge .... HYSTEROGRAPHIUM
  - 2. Ascostromata conchate, erect .... OSTREOLA

V. Ascospores filiform, ascostromata conchate .... LOPHIUM

Farlowiella: The species of this monotypic genus have elongate superficial ascostromata with longitudinal slits and one-celled papillate ascospores. A conidial state, *Monotospora megalospora,* has been reported by Mason (**1914**).

Glonium: Glonium has ellipsoid to elongate or branched ascostromata, often clustered upon a subiculum. The species without a subiculum have been placed in a genus *Psiloglonium* by some authors. The ascospores are two-celled and hyaline. Several species have pycnidial states of the form genera *Sphaeronaema* or *Plenodomus. Glonium clavispora* has a *Sporidesmium* state with irregular, brown, muriform conidia borne singly on slender conidiophores. Lohman (**1931, 1937**) has reported upon the life histories of this genus. (The genus *Lophiosphaera,* in the Lophiostomataceae, differs only in the ascostromata being sunken in the substrate.)

The development of *Glonium stellatum* was studied by Luttrell (**1953**). In this species a felt-like subiculum is formed on the surface of decaying wood. Pycnidia and ascocarps arise on this subiculum. The conidial state, *Sphaeronaema biffusoideum,* arises as globose stromata, on this subiculum, which eventually elongate to form ostioles. The cells in the interior of this stroma elongate to form conidiophores which abstrict small ellipsoid, hyaline conidia from their tips. The ascocarps are elongate or dichotomously branched and arranged in radiate rows upon the subiculum. They originate as globose stromata with a central hyaline parenchyma and an outer pigmented layer of hyphae. Within this stroma a deep staining coil arises which is probably the archicarp. Its cells soon become binucleate and then multinucleate. Ascogenous hyphae arise as outgrowths from these cells. At this time the stroma expands laterally to become ellipsoid and then elongate in shape. As the ascocarp elongates, the ascogenous hyphae grow out from the mass of ascogenous cells, keeping pace with the stretching of the stroma. At the same time, the parenchyma about the ascogenous hyphae disintegrates and gives rise to a cavity. The upper cells of the stroma remain meristematic and filiform hyphae grow downwards from these cells forming a loose weft of "pseudoparaphyses." The asci probably arise from croziers on the continuous layer of ascogenous hyphae in the base of the cavity, and grow upwards between the pseudoparaphyses. A slit-like opening of the ascocarp originates merely as a cleft in the surface layer of dark parenchyma. The asci are thick-walled at first but become thin- and double-walled at maturity. Upon spore ejection the outer thin wall splits near the apex and the thick inner wall becomes gelatinous and swells, pushing outward and carrying the ascospores with it. This is typical of the bitunicate and ascolocular type of development found in the Dothideales. The presence of true pseudoparaphyses is somewhat anomalous but would probably indicate a relationship with the Pleosporales near the Lophiostomataceae. It indicates also the sort of variation which confused the original attempt to separate the types of centrum development.

Gloniella: This genus has elongate ascostromata and hyaline ascospores with more than two cells. Some species have ascospores which become brown when fully mature and have been, therefore, transferred to *Hysterium. Hysterium karstenii* (*Gloniella ambigua*), with a *Sporidesmium* state, is such a species.

Hysterium: This genus is similar to *Gloniella* but the ascospores are many-celled and brown. *Hysteropycnis* is given as the conidial state of *Hysterium pulicare* by Hilitzer (**1929**), and Lohman (**1931**) finds an *Aposphaeria* type of pycnidium in *Hysterium minutum. Hysterium hyalinum* (Lohman, **1934**) produces conidia of a *Sporidesmium* with clavate, brown, muriform conidia on short stalks, and also pycnidia. Lohman (**1933**) also reported *Septonema spiloneum* and a pycnidial state for *Hysterium insidens.* This *Septonema* was transferred to *Sirodesmium* *granulosum* by Hughes (**1952**) because the septate brown conidia are produced in chains from compacted conidiophores, basipetally, by the division of a basal cell, so that the apical conidia are the most mature, whereas in *Septonema* they are produced acropetally with the oldest spores at the base of the chain *Glonium, Gloniella* and *Hysterium* grade off one into the other and the conidial states are similar.

Gloniopsis: *Gloniopsis* has elongate ascocarps and muriform ascospores which are hyaline. Lohman (**1933**) reports pycnidia of the *Aposphaeria* type for *G. gerardiana*

Hysterographium: This genus also has ellipsoid to elongate ascocarps, but the ascospores are muriform and brown. A number of species of *Hysterographium* have been reported as having pycnidial states. Hilitzer (**1929**) placed these in *Hysteropycnis.* Lohman (**1933**) reported superficial pycnidial forms as in *Aposphaeria* and sunken ones as in *Dendrophoma.*

Hysterocarina: Zogg (**1949**) erected this genus for a species similar to *Hysterographium* but with ascocarps having a keel-like ridge along the slit-like opening. The base of the ascocarp is sunken in the wood

Bulliardella: In this genus the ascostroma is conchate, shaped like a clam shell standing on edge on the substrate. It has a thin membranous wall and an elongate slit along the upper edge. The ascospores are two-celled and brown. Zogg (**1960**) gave *Bulliardiella* as a synonym of the earlier genus name *Actidium*, which he uses in his monograph (**1962**). *Actidium* has lobed or stellate ascocarps, but some species of *Bulliardiella* show branching or triradiate forms.

Mytilidion: This genus has conchate ascocarps as in *Bulliardiella*, but the ascospores are many-celled and brown. Several species of *Mytilidion* have produced pycnidia (*Aposphaeria, Pyrenochaeta,* and *Sclerochaeta*) in culture (Lohman, **1933a**). *Mytilidion scolecosporum* has produced *Septonema toruloideum* (Lohman, **1933**). Certain species of *Mytilidion* with long-cylindric, septate, biseriate ascospores which have a ratio of length to width of twenty:one, or about half the length of the ascus, are intermediate between *Mytillidion* and *Lophium* and have been placed by Lohman (**1932**) in a subgenus *Lophiopsis.*

Ostreion: This a genus which differs from *Mytilidion* only in the fusoid character of the ascospores.

Lophium: Lophium has shell-shaped ascostromata and filiform, septate, ascospores lying parallel to one another the full length of the ascus. Lohman (**1933**) found two conidial states of *L. mytilinum*. One consists of spore balls of *Papulospora mytilina*, which originate as coiled hyphae that develop into multicellular heads. The other is a pycnidial state of the form genus *Dothiorella*, consisting of clusters of blackened, immersed pycnidia with erumpent ostioles and forming ellipsoid, one-celled, hyaline conidia. Zogg (**1962**) has placed species with flattened, "hatchet-shaped" ascocarps in a separate genus *Glyphium*.

EUASCOMYCETES

(Ascohymeniales, Unitunicatae)

The Euascomycetes include those species which have a true perithecium. Such an ascocarp arises as an archicarp, usually a coiled hypha, upon the vegetative mycelium. The perithecial wall arises by the proliferation of hyphae from the base of the archicarp or from neighboring hyphae. Any stroma that is present arises by the proliferation of hyphae that are distinct from those of the perithecium. The asci are thin-walled, unitunicate, usually with a thickened apical wall, and often including various types of modifications known as an "apical apparatus." The group includes a wide variety of perithecial and stromatic configurations and ascus structure, which serve for the separation of taxa.

The Plectascales include simple forms without any preformed opening to the fruit body, or "cleistothecium." They are mostly saprophytic, occurring on soil, dung or decaying vegetable remains. Most of the parasitic species are found on animals or on man. Such forms as the Gymnoascales can be easily derived from the Endomycetales of the Protoascomycetes.

The simpler forms of the Plectascales have globose asci which are distributed throughout the cleistothecium centrum. Other members of this group show the development of clavate to cylindric asci and a clustering of the asci into a central or basal fascicle. Such species may represent the simpler forms of some of the more advanced evolutionary species of the unitunicate series. The Melanosporales show this advanced arrangement and the formation of an ostiolar neck.

From this point onward there seems to be an increasing number of evolutionary series, as in the Hypocreales, Xylariales, Diaporthales, Allantospheriales and many families which do not fall into concise ordinal groupings.

In the older systems there was a separation of the simple from the stromatic Sphaeriales. It has become increasingly evident, however, that the formation of a stroma has occurred in many evolutionary series, containing both simple and stromatic forms.

The stroma may take on different appearances. In the Hypocreales and most of the Clavicipitales, the stroma is bright colored, and the perithecia arise in or on the preformed, usually erumpent stroma. In most other stromata, two elements may be recognized. These stromatic species are mostly wood inhabiting. The "ectostroma" is the first to be formed and is usually on or just beneath the surface of the substratum. It usually bears the conidial state and may serve as a mechanical tissue to rupture the overlying periderm on corticated stems. The "entostroma" usually develops within the substratum beneath the ectostroma. It is in this entostroma that the perithecial initials originate and develop (Wehmeyer, 1926).

In the Xylariaceae, the stromata are well developed and usually erumpent, and may be considered an entostroma. Conidial states may be formed on its surface, but the perithecia arise within or upon this stroma. At maturity there is a differentiated surface, cortical, black or colored layer which can be looked upon as an ectostroma. The remaining stromatic forms are immersed or erumpent upon woody substrata. In these species three types of stromata may be recognized. In the "eutypoid" type, the stroma is effuse over a more or less wide area and the perithecia are scattered singly or in loose groups and the ostiolar necks are separately erumpent. The entostroma is not always obvious as it may consist of a proliferation of fungus hyphae between the cells of the host substratum. It may be detected by the blackening of the host substratum or by the presence of an outlining blackened zone which may penetrate into the bark or the wood. This blackened zone consists of dark-walled hyphae which delimit the stromatic area from the disintegrating action of the nutrient hyphae within the bark. It corresponds to the blackened surface of stromata exposed to the air, and occurs in many types of stromata.

The "diatrypoid" type of stroma is disc-shaped or more widely effused and the perithecia are also separately erumpent, but they are imbedded within a strongly developed entostroma which becomes erumpent and throws off any ectostromatic developemt which has taken place. The surface of this entostroma has a differentiated colored layer and the stroma may be outlined by a blackened zone in the substrate.

The "valsoid" type of stroma consists, in most cases, of an ectostroma which develops as a conic or disc-shaped stroma beneath the periderm and serves to rupture the overlying periderm through which it is erumpent as a small disc. Conidial states may occur in locules within this ectostroma. When perithecia are formed they occur in a cluster in an entostroma which is formed beneath the ectostroma. The perithecial necks have an apical meristematic growing point and grow upward through the ectostromatic disc. There may or may not be a blackened zone partially or entirely surrounding one or more such valsoid clusters of perithecia.

The simple unitunicate species may have scattered, gregarious or crowded individual perithecia either superficial or immersed within the substratum. They may show various indications of the beginning of stromatic formation. The superficial species may be seated on a subiculum of interwoven colored hyphae or upon a thin layer of parenchyma. In the immersed species there may be a development of hyphae about the ostiole discoloring the epidermis or surface host tissue to form what is known as a "clypeus."

PLECTASCALES

The Plectascales comprise a group in which the fruit body is entirely enclosed within a wall with no differentiation of an opening for the dispersal of the ascospores. The asci are for the most part globose, saccate or broadly clavate, and are dispersed in an irregular manner within the wall of the ascocarp. Such a fruit body is known as a "cleistothecium." The order includes several families which are related by their morphology but not necessarily phylogenetically. The members of the Plectascales are widely distributed on a variety of substrata. They are found as saprophytes on soil, dung, vegetable remains, and on bones and feathers. Some occur as parasites on the skin and hair of animals and man. Many species have various types of hyphomycetous conidial states, some of which, as *Aspergillus* and *Penicillium*, are common laboratory contaminants and are destructive molds and rots of fruits and vegetables. The following families are included here:

I. Fruit body stalked
    A. Sterile veins between columns of evanescent asci; open above TRICHOCOMACEAE
    B. Not as above; asci scattered throughout a globose to pulvinate head .... ONYGENACEAE
II. Fruit body not stalked
    A. Wall of fruit body a loose weft, or peridium, of modified hyphae .... GYMNOASCACEAE
    B. Fruit body with a firm wall
        1. Hypogeous (subterranean) fungi .... ELAPHOMYCETACEAE
            a. Core corky or woody; spores hyaline .... Mesopheleae
            b. Core cottony or fleshy; spores brown .... Elaphomyceteae
        2. Not hypogeous
            a. Ascocarp ascohymenial, arising from a coiled archicarp
                i. Spores with a small lower appendage-like cell. TRIPTEROSPORACEAE
                ii. Spores not so appendaged......................... EUROTIACEAE
            b. Ascocarp arising as a parenchymatic stroma .... PHAEOTRICHACEAE

GYMNOASCACEAE:

This family includes members which have a loose network or "peridium" around the clustered asci. The peridial hyphae may be simple vegetative hyphae or may be modified by thickened walls or various types of appendages. They attack substrata with the animal protein keratin, such as hair, feathers, skin, bone, etc. They are also found on dung and in soil. The fungi causing dermatomycoses of humans and higher animals are believed to be the imperfect state of some of these fungi. They cause such diseases as athletes' foot, barbers' itch, ringworm, tinea capitis, psoriasis, etc.

The following key is taken from those of Kuehn (**1958**) and R.K. Benjamin (**1956**).

I. Peridium absent; asci in loose groups within the vegetative hypae
  A. With a *Paecilomyces* imperfect state. . . . . BYSSOCHLAMYS
  B. With no imperfect state . . . . PSEUDEUROTIUM
II. Peridial hyphae thin-walled similar to vegetative hyphae
  A. Ascospores light colored . . . . ARACHNIOTUS
  B. Ascospores brown to brown violet . . . . AMAURASCUS
III. Peridial hyphae thick-walled, differentiated from the vegetative hyphae
  A. Peridium merely thick-walled not otherwise differentiated; without spines. . . . . .
  . . . . PSEUDOGYMNOASCUS
  B. Peridium variously differentiated
    1. Peridium anastomosed to form a net; with short spine-like appendages . . . . .
    . . . . GYMNOASCUS
    2. Peridium forming bramble-like aggregates, with recurved spines; some internal coils . . . . EIDAMELLA
    3. Peridium composed of constricted asperulate cells; appendages few . . . .
    . . . . ARTHRODERMA
    4. Peridium of disarticulating irregular cells; with numerous internal spiral appendages. . . . . SHANORELLA
    5. Peridium with appendages straight or uncinate. . . . . MYXOTRICHUM
    6. Appendages septate with recurved tooth from each cell. . . . . CTENOMYCES

Byssochlamys is an interesting genus because it represents a link between the *Endomycetales* and the *Gymnoascaceae*, which in turn can be looked upon as the origin of the ascohymenial or Euascomycete group.

The fruiting body originates as a coiled hypha from which ascogenous hyphae arise; asci are formed from croziers in a loose cluster lying in a weft of vegetative mycelium without any differentiated peridium. It is common in the spoilage of fruits and preserves. A conidial state of the form genus *Paecilomyces* is formed, which bears phialid-like branches as in *Penicillium*, relating the genus to the Eurotiaceae. Some species also form ovoid aleuriospores similar to those found in *Eremascus*. The ascospores are globose, smooth and borne in globose asci.

Pseudoarachniotus of Kuehn (**1958**) is similar to *Byssochlamys* in having a loose vegetative peridium of undifferentiated hyphae, but it has no *Paecilomyces* conidial state. Kuehn would place *Byssochlamys* in the Eurotiaceae because of the *Paecilomyces* conidial state.

Arachniotus has been found on dung. The fruit body originates from two coiled gametic hyphae, one of which becomes septate, including a number of ascogenous cells which give rise to ascogenous hyphae. According to DeLamater (**1937**) each ascogenous cell is binucleate and buds out as a crozier which results in a cluster of asci. Undifferentiated vegetative hyphae sometimes bearing coiled hyphae, surround the clustered asci. The ascospores are hyaline to yellowish in mass, globose and smooth. Clavate aleuriospores are also formed.

Amaurascus has a development and structure very similar to *Arachniotus,* but the ascospores are brown to violet-brown.

Gymnoascus is the best known of the genera, but its limits have been confused. Baranetsky (**1872**) was the first to describe ascocarp formation in the Gymnoascaceae. In *Gymnoascus* there are two gametic hyphae. The "antheridial" branch becomes clavate and cuts off a terminal cell. The "ascogonial" branch coils about the antheridial one and becomes septate. Some writers claim there is a fusion between the two; others fail to see any fusion. The septate cells of the ascogonial branch give off ascogenous hyphae which become septate into binucleate cells, each of which give rise directly to a globose ascus. The peridial branches grow out from the base of the copulating hyphae and encircle the asci with a network of branching hyphae, the tips of which are spine-like and have spine-like or recurved branches. *Gymnoascus* has been found on dung and feathers.

Pseudogymnoascus is similar to *Gymnoascus*, but the network of thickened hyphae does not have lateral spines.

Eidamella has been reported from the skin of a dog (Matruchot and Dassonville, **1901**). It is also found as a saprophyte on paper, rags, etc. Development is similar to that of *Gymnoascus*, but the peridial hyphae form a brush-like mass of blackish, thick-walled hyphae with recurved spine-like branches, but not in a true net. Small coiled, hyaline hyphae are found on the inner branches. The ascospores are ellipsoid.

Arthroderma has peridial hyphae composed of nearly hyaline, thick-walled asperulate cells which are constricted at the septum to form chains of globose cells. It also forms clavate aleuriospores. It is found on feathers and old horn.

Shanorella was found on rabbit fur, chicken feathers, etc. (R.K. Benjamin, **1956**). The peridium is made up of irregularly shaped cells which are pale yellow and fall apart at maturity. Thin-walled hyaline coiled hyphae are formed as lateral outgrowths. The ascospores are flattened, disc-shaped and bright yellow.

Myxotrichum has a network of thick-walled, septate hyphae with spine-like branches, and long appendages with circinate tips on the surface. It forms chains of clavate aleuriospores on the vegetative hyphae.

Ctenomyces occurs on feathers. The peridium consists of a thin inner layer of interwoven thin-walled hyphae and a lax outer network of thick-walled, orange-brown hyphae which give rise to lateral curved appendages made up of thick-walled asperulate cells, each of which gives rise to a short, stout, recurved, lateral branch.

ONYGENACEAE

This family has stalked ascocarps with numerous asci in a terminal swollen head. It includes several probably unrelated genera.

Onygena: Onygena equina is found on the horns and hooves of horses and cattle. It has a white to cream colored stalk, the hyphae of which are spread out at the apex to form a globose head. Numerous asci are formed in this globose head. *Onygena cervina* occurs on bird feathers.

Chadefaudiella: This genus is found on gazelle dung. The fruit body has a basal stalk and an ovoid head with a parenchymatic wall. There is a basal hymenium of evanescent asci, leaving a mass of fusoid ascospores intermingled with a network of capillitial threads. Faurel and Shotter (**1959**) placed this genus in a separate family, Chadefaudiellacae, because of the capillitial threads and walled ascocarp.

Rollandina: Found on plant debris, this is another isolated form. The fruiting structure consists of an elongate stipe with a somewhat lobate head. This head is composed of branched, septate, asperulate hyphae, within which there occur numerous small, white, glomerules, each consisting of a mass of ovoid asci containing eight disc-like ascospores. Large yellow-brown five-septate, fusoid conidia are formed laterally on the hyphae of the head.

Roesleria: This is a genus which Arnaud placed in his Caliciaceae, but in structure it seems related to this family. It causes a root rot of grape and other plants and is found underground. It has a short stalk and a globose head more or less enclosed in peridial hyphae. The asci are formed in the head and deliquesce to leave a mass of ascospores, which are probably distributed by insects, earthworms, etc.

Penicilliopsis: This genus has a stalked fruit body with an irregular pulvinate head. The asci are borne in irregular masses within cavities of this head (Solms-Laubach, **1887**).

The form genera *Microsporum, Epidermophyton* and *Trichophyton* are the causes of the skin and hair diseases known as dermatomycoses. They have been found to be the conidial states of members of the Gymnoascaceae in a number of cases. The genus *Nannizzia* was erected for the perfect stages of *Microsporum*.
The following connections have bee made:

Microsporum gypseum — Nannizzia incurvata
Keratinomyces ajelloi — Arthroderma uncinata
Trichophyton terrestre — Arthroderma quadrifidum
Microsporum nanum — Nannizzia obtusa
Microsporum cookei — Nannizzia cajetana
Microsporum breuseghemii — Nannizzia grubyia
Trichophyton gypseum var. *asteroides — Ctenomyces trichophyticus*

A full account of these fungi and the diseases caused by them can be found in "A Manual of Clinical Mycology" (Conant, et al., **1954**).

EUROTIACEAE:

The members of this family differ from those of the Gymnoascaceae in the presence of a firm outer wall. It is probably a heterogeneous group with cleistothecial forms of various other lines of descent. Many of the genera of the family have been separated upon the basis of their conidial states.

Cain (**1956b**) argues that the species of Pyrenomycetes with ostioles or ostiolar necks are adapted to distribution of their spores by air currents. The cleistothecial forms, he believes, have been derived from the ostiolate forms by adapting themselves to situations where air currents do not obtain, such as dry situations, vegetable debris, soil, etc. They disperse their spores in other ways. Some have gelatinous spores which are dispersed by insects or other animals; others have persistent ascocarps which rupture and disperse spores over a long period. He points out that the Plectascales are a heterogeneous group with species showing derivation from various ostiolate taxa. There is good evidence that Cain is correct in saying that there are species related to other higher taxa; but whether they have been derived from ostiolate species or vice versa is open to argument. Such relationships will be indicated in later discussion. Inasmuch as these relationships have not been fully worked out, and for convenience, all of these "unrelated" genera will be included under the Eurotiaceae. Their probable relationship to other taxa will be discussed under the respective genera.

Ascocarp formation is of the ascohymenial type in most genera, but varies in minor details from species to species. It usually begins as one or two gametic hyphae which coil about one another. There may or may not be a pore and fusion between such gametic hyphae. Nuclear fusion in the ascogonial cells has been reported in a few cases, but has not been actually seen in most species. At any rate, ascogenous hyphae arise from the ascogonial cells and proliferate to eventually give rise to binucleate cells which, in turn, give rise to asci either with or without crozier formation. Nevertheless, there are some species which have an ascolocular type of development, by the division of modified vegetative hyphae to form a stromatic ascocarp, which indicates the heterogeneity of this group.

I. Cleistothecial wall thin, filled with loose ascospores from evanescent asci. . MONASCUS
II. Not as above
- (I). Spores 1-celled
  - A. Spores hyaline
    - 1. Fruit body light colored
      - a. With *Aspergillus* conidial state
        - i. Wall of a single row of parenchyma; spores with furrow . EUROTIUM
        - ii. Wall of interwoven hyphae; spores with frills or bands . . SARTORYA
      - b. With *Penicillium* as conidial state
        - i. Wall of interwoven hyphae; asci in chains; spores spiny or frilled. . ... TALAROMYCES
        - ii. Wall of parenchyma; asci on side branches . . . . CARPENTELES
      - c. Neither *Penicillium* nor *Aspergillus* states; *Eurotium*-like, yellow-brown; conidia clavate to ovoid. . . . . ANIXIOPSIS
    - 2. Fruit body dark colored, brown or black
      - a. Spores globoid . . . . MYCOGALA
      - b. Spores with wing-like appendages . . . . SAMAROSPORA
      - c. Spores with 3-cornered wings; *Leptographium* conidial state . . . . EUROPHIUM
    - 3. Fruit body deep purple; wall fragmenting into plates . . . FRAGOSPHAERIA
  - B. Spores purple-red, winged and grooved . . . . EMERICELLA
  - C. Spores brown to black; fruit body usually dark
    - 1. Fruit body setose
      - a. Hairs straight . . . . CHAETOTHECA
      - b. Hairs spiral or circinate. . . . . KERNIA .... (MAGNUSIA)
      - c. Hairs simply curled at apex; asci evanescent. . . . . CHAETOMIDIUM
    - 2. Fruit body not setose
      - a. Wall breaking up into plates; spores fusoid. . . . . CEPHALOTHECA
      - b. Wall not breaking up as above
        - i. Spores with 2 wing-like flanges; *Cephalosporium* conidial state . . . . EMERICELLOPSIS .... (SATURNOMYCES)
        - ii. Spores with spiral ridges. . . . . WESTERDYKELLA

iii. Spores cuboid. . . . . PHAEIDIUM
iv. Spores ellipsoid to globose
(a). Spores with germ pores
(1). Spore wall smooth. . . . . THIELAVIA
(2). Spore wall with circular pits . . . . ANIXIELLA
(b). Spores without germ pores
(1). Conidial state a pycnidium . . . . PYCNIDIOPHORA
(2). Conidial state *Sporotrichum* . . . . PSEUDEUROTIUM
(3). Conidial state *Monosporium* . . . . ALLESCHERIA
(II). Spores 2-celled; mostly dark colored
A. Cleistothecia setose with long hairs
1. Spores with a slimy appendage. . . . . ZOPFIELLA
2. Lower cell of spore hyaline, as an appendage
a. Spores smooth . . . . TRIPTEROSPORA
b. Spores with circular pits in wall. . . . . DIPLOGELASINOSPORA
3. Spores without an appendage, with terminal germ pores. PHAEOTRICHUM
B. Cleistothecia not hairy
1. Spores large, 40 μ or more . . . . ZOPFIA
2. Spores smaller
a. Fruit body dark colored. . . . . TESTUDINA
b. Fruit body light colored; spores smooth, hyaline becoming orange-red . . . . HELEOCOCCUM
(III). Spores 1-celled, but combining in chains of two to eight. . . . . SYNAPTOSPORA
(IV). Spores 4-celled, fragmenting into 1-celled fragments. . . . . PREUSSIA

Monascus: This genus is usually placed in a separate family, the Monascaceae. Earlier writers interpreted its cleistothecium as a polysporous ascus or a sporangium, related to the Phycomycetes. The cleistothecium arises, however, as two gametic branches (Barker, **1905**; Kuyper, **1905**; Olive, **1905**) as in most Ascomycetes. The ascogonial branch arises just below the septum of a terminal or intercalary cell. It grows up parallel to the adjacent cell and bends it aside at a right angle, sometimes coiling about it. The side branch is cut off by a septum and fuses with the other cell by a wide pore. Both gametic hyphae are multinucleate. The ascogonial branch divides into two cells by a septum. The penultimate cell swells to form the "central cell." Investing hyphae grow up from below and surround this central cell which becomes large and refractive. A knot of ascogenous hyphae grows out from this central cell. These ascogenous hyphae break up into globose cells which become the asci. The walls of the asci and the ascogenous cells break down and the central cell within the hyphal wall becomes filled with loose ascospores.

Monascus ruber causes a reddish-purple discoloration and is used in the Orient to make pigmented red rice or Samsu, an alcoholic drink called Beni Koji and a pigment used in cooking called L'Ang Quoc. *Monascus ruber* and *M. purpurea* also cause a mold of rubber.

Talaromyces and *Carpenteles* have a *Penicillium* conidial state. The wall of *Talaromyces* is formed of interwoven hyphae and the asci are globose and borne in chains. In *Carpenteles* the wall is parenchymatic and the globose asci are borne on short side branches.

Penicillium is a most widespread genus with hundreds of species which are difficult to separate. The group has been monographed by Raper and Thom (**1949**). In *Penicillium* the conidiophores arise as upright branches directly from the vegetative hyphae. At their apex they become branched one or more times in a finger-like manner into "metulae," each of which bears one or more "phialides." Phialides are bottle-shaped conidiophore branches with a constricted neck. The apical cell wall of this neck breaks open and conidial masses of protoplasm are cut off from the cytoplasm of the phialid and extruded as endogenous conidia.

The species of *Penicillium* grow on all sorts of substrata, especially those with abundant carbohydrates. *Penicillium italicum, P. expansum,* and *P. digitatum* cause green and blue molds of citrus fruits. Other species cause rots and molds of jams, jellies, meats and bread. *Penicillium notatum* and *P. chrysogenum* are the source of penicillin. By the development of favorable strains the yield of penicillin has been increased one hundred fold or more.

Pontecorvo and Roper (**1952, 1956**) claim that mutations may take place in these asexual fungi, and fusion of nuclei and segregation of genetic characters may occur. This occurrence has been given the of "parasexuality."

Penicillium camembertii and *P. roguefortii* are used in the production of cheeses of these names. Many species are of technical importance because of their ability to ferment starches, sugars and tannins to produce citric, gallic, oxalic acids and various alcohols.

The ascus state is not kown in all species of *Penicillium* and this state often takes place only under special cultural or environmental conditions. Emmons (**1935**) recogized two groups of *Penicillium* perfect states which were later given the names of *Talaromyces* and *Carpenteles* by C.R. Benjamin (**1955**). In *Talaromyces* the cleistothecium arises as various types of copulating branches, one of which coils about the other, as described for *Gymnoascus*. The asci are formed in chains directly from the cells of the ascogenous hyphae, without crozier formation. The cleistothecial wall is filamentous and the spores are spherical with a roughened wall.

In *Carpenteles*, the cleistothecium arises as a dendroid mass of branching hyphae, which by anastomoses and septation form a parenchymatic ascocarp or "sclerotium," before the asci appear. The asci are formed on short side branches of the ascogenous hyphae in a central cavity of the sclerotium. The globose ascospores commonly have an equatorial groove.

Eurotium, Sartorya and *Emericella* have the form genus *Aspergillus* as the conidial state. In *Aspergillus* the upright conidiophore arises from a "foot cell" cut off by septa in a vegetative hyphae. It forms a spherical to clavate swollen vesicle and form phialides at their apex, which produce endogenous conidia as in *Penicillium*. In some species the branches or metulae branch and form a second or third series. Such species have been placed in the genus *Sterigmatocystis.*

The numerous species of *Aspergillus* are just as wide-spread and difficult of determination as those of *Penicillium*. They have been monographed by Thom and Church (**1926**) and Raper and Thom (**1949**), who have sorted them out into "groups" of overlapping species and varieties. *Aspergillus* species are more common in warm climates and are common in soil. The antifungotics fumagacin and clavacin are formed by *A. fumigatus* and *A. clavatus*. *A. oryzae* is used in the production of the rice liquor Sake. It is associated with a yeast. The *Aspergillus* enzymes break down the starch into simple sugars which are then fermented by the yeast. Soya sauce is produced by this same species by its action on the proteins of the soy bean. *Aspergillus flavus* causes a disease of honey bees and *A. fumigatus* a disease of birds and sometimes of humans.

Cleistothecial formation is favored by abundant carbohydrates. Cleistothecia originate as coiled gametic hyphae (often only one branch). No fusion occurs in most species. The ascospores consist of two halves like those of a clam shell. It they fit evenly, a discoid spore is formed, but usually there is an equatorial furrow where they meet. Other spores have various types of elaborate flanges formed at the edge of the furrow. *Eurotium* has an ascocarp wall consisting of a single row of sclerotic parenchyma cells. The ascospores, for the most part have an equatorial furrow. This type of ascocarp has the conidia belonging to the *A. glaucus* group of species.

Sartorya has an ascocarp wall of interwoven hyphae. The growth is indeterminate, the ascocarp increasing in size over a long period of time. The ascospores have frills or bands as well as an equatorial furrow. This type of ascocarp has conidia in the *A. fischeri* group.

Emericella species produce conidia of the *A. nidulans* group of *Aspergillus*. The ascocarp wall is of interwoven hyphae, but has an envelope of enlarged, inflated "Hulle" cells. The ascospores are purplish red in color and have elaborate frills or dentate wings on the edges of the equatorial furrow.

A number of form genera have been segregated from those having a *Penicillium* or *Aspergillus* state, as follows: *Aspergilloides* Dierck was erected for those species with a monoverticillate *Penicillium*, with only one series of metulae.

Microaspergillus Wehmer was erected for monoverticillate Penicillia which have a foot cell as in *Aspergillus*.

Aspergillopsis Sopp was erected for Penicillia with a clavate swollen apex, from which the metulae arise.
Citromyces Wehmer included penicillia which produce citric acid.

Synpenicillium Const. was used for ropy Penicillia with simple lateral heads. Saccardo put these in *Coremium* and Thom put them in *Scopulariopsis*.

Clonostachys Corda was erected for those having conidiophores verticallately branched in two tiers and spores arranged spirally. This was put in *Gliocladium* by Raper.

Sterigmatocystis for Aspergilli with sterigmata in more than one series; *Diplostephanus* was used for the ascosporous states of *Sterigmatocystis*.

Aspergillopsis Speg. included the black spored Aspergilli.

Thermoascus was used for species with a *Dactylomyces* conidial state.

Anixiopsis is similar to *Eurotium*, but has no *Aspergillus* state. Some species have chains of swollen cells as conidia. The ascospores are warty and the asci lie in a central core of hyaline parenchyma (Stolk, **1955a**).

Mycogala has dark colored cleistothecia and hyaline globose spores.

Samarospora has dark colored fruit bodies and hyaline spores with a wing-like appendage.

Europhium (Parker, **1957**) has dark leathery fruit bodies. The asci are evanescent and the spores are hyaline with a three-pronged wing and are imbedded in a gelatinous matrix. The conidial state is a *Leptographium*. These characteristics place it near *Grossmania* in the Ceratocystaceae, but it lacks the ostiole of that genus. Like the latter genus, it was also isolated from blue-stained pine wood.

Fragosphaeria has dark purple cleistothecia in which the wall breaks up into angular plates at maturity, much as in *Cephalotheca*. The spores are hyaline in *Fragosphaeria*, dark in *Cephalotheca*.

Chaetotheca has a fruit body with long straight hairs, and dark colored spores.

Kernia (*Magnusia*) has long spiral or uncinate appendages on the fruit body, surrounded by light colored flexuous hairs in some species.

Chaetomidium has short curved appendages on the fruit body. The asci are evanescent and the spores lie free in the cleistothecium. The spores have a germ pore at the pointed end. It differs chiefly from *Chaetomium* in the lack of an ostiole and could be placed in the Chaetomiaceae.

Emericellopsis (Stolk, **1955**) [*Saturnomyces* (Cain, **1956**)] has dark colored cleistothecia and dark gray to black ellipsoid ascospores with three wing-like appendages. The conidial state is a *Cephalosporium*.

Westerdykella has dark colored cleistothecia which arise by the division of vegetative cells to form a parenchymatic stroma, not by a coil as in a true perithecium. The asci are in a fascicle within a hyaline parenchyma matrix. The origin of the fruit body is similar to that found in the Phaeotrichaceae (see below) and the ascus arrangement is intermediate between the loose mass of asci in the Eurotiaceae and the hymenial arrangement of the Melanosporales. Similar arrangements are found in *Anixiopsis, Pseudeurotium, Thielavia* etc. The ascospores of *Westerdykella* are globose, thirty-two in an ascus and ornamented with spiral bands.

Phaeidium has ascospores which become cuboid.

Thielavia has dark brown ellipsoid to angular spores with one or two germ pores at the ends. Cain (**1961**) placed it in his family Tripterosporaceae. *Thielavia basicola* is found on the roots of many plants and causes a black root rot of tobacco. It is usually accompanied by two conidial states, *Thielaviopsis* and a *Torula*-like chain of chlamydospores. These are not genetically connected to the *Thielavia* but are necessary as a stimulant to abundant formation of cleistothecia. In *Thielavia sepedonium* there is a *Sepedonium* state. Doguet (**1956**) found that the cleistothecium of *T. terricola* arises as a coiled hypha, around which enveloping hyphae form a wall. The tip cell of this coil cuts off a single binucleate cell from which one to three chains of five to seven uninucleate cells arise. The apical cell of each chain gives rise to a crozier which forms a fascicle of asci by repetition. The asci break down in a cavity in the central parenchyma tissue and lie in an irregular fascicle. The development is similar to that found in *Melanospora*, except for the lack of an ostiole.

Anixiella has one-celled brown ascospores whose walls are covered with small, hyaline, circular pits. They have a distinct germ pore at each end. This wall ornamentation is similar to that found in *Gelasinospora*, to which this genus is related, differing only in the lack of an ostiole. The asci have a refractive ring in the thickened apical wall, similar to that found in the Sordariaceae.

Pycnidiophora was erected by Clum (**1955**) for a species with cleistothecia containing thin-walled globose asci, but which had a pycnidial conidial state. The one species, *P. dispersa,* was given as having asci with thirty-two ascospores. Cain (**1961**) placed this species in *Preussia* believing that the thirty-two spores in the ascus were derived from the fragmentation of eight four-celled ascospores. Kowalski (**1964**) in a study of the development of the fruit body, found that it arose by the division of modified cells of the vegetative hyphae to form a parenchymatic stroma. This he said would place the genus in the Phaeotrichaceae. He found that the ascospores were not four-celled but were cut out as thirty-two one-celled ascospores. Therefore the species was excluded from *Preussia.*

Pseudeurotium (Van Beyma, **1937**) has light olive to brown ascospores, and differs from *Thielavia* chiefly in the lack of pores in the ascospores. *Sporotrichum* has been found to be the conidial state of several species. In *P. ovalis,* the cleistothecium arises as a row of swollen vegetative hyphae which divide in all directions to form a parenchymatic stroma. Ascogenous hyphae arise from undifferentiated cells in the interior of this stroma. This is typical of the Ascoloculares and would place this species in the Phaeotrichaceae if one adhered strictly to the unitunicate-bitunicate theory. However, the asci are globose to stout clavate, up to thirty-two spored and do not show a bitunicate type of emission.

Allescheria differs from *Pseudeurotium* chiefly in the conidial state which is a *Monosporium*. *Allescheria boydii* was isolated by Shear and Dodge (**1922**) from a mycetoma of the foot. Its conidial state is *Monosporium apiosporium*. The spores are light brown, globose to ovoid.

Zopfia has two-celled brown ascospores which are very large (40 μ or more). The cleistothecia are not setose.

Zopfiella has two-celled brown spores with a slimy appendage. The cleistothecia have long filiform hairs.

Diplogelasinospora has ascospores which are two-celled with the lower cell remaining as a hyaline appendage, as in *Tripterospora*. It differs in having circular pits in the ascospore wall as in *Gelasinospora* and *Anixiella*, which relates it to the Sordariaceae, but it lacks an ostiole.

Testudina is similar to *Zopfia*, but has smaller two-celled dark ascospores.

Helecoccum has a light colored fruit body and ascospores which are smooth and hyaline to orange-red.

TRIPTEROSPORACEAE:

This family was erected for the genus *Tripterospora* by Cain (**1956**). It shows many characters found in the Sordariaceae. The spores are one-celled at first but cut off a short or slender lower cell which remains as a hyaline appendage whereas the upper cell becomes dark brown. Cain says the asci sometimes have a ring in the apex, but they are clavate, evanescent and do not lie parallel in a hymenium.

PHAEOTRICHACEAE:

Cain (**1956a**) erected this family for the genus *Phaeotrichum*. The cleistothecia bear straight or uncinate appendages. The spores are dark brown, two-celled with a hyaline germ pore at each end. Barr (**1956**) has shown that the cleistothecium arises from vegetative hyphae by division of the cells to form a parenchymatic stroma, within which ascogenous hyphae arise, in turn giving rise to the asci. This is the type of development found in the Ascoloculares and, therefore, deserves a separate family. The asci are clavate but not typically bitunicate. They have no thickened apical wall, and are evanescent within the fruit body cavity. *Preussia* is a puzzling genus. Cain (**1961**) described the cleistothecium as non-ostiolate and arising by the division of vegetative cells to form a parenchymatic stroma. The asci may be globose and short-stalked and arranged in an irregular fashion within the cleistothecium, or cylindric and long-stalked, and then more or less parallel in a hymenium with accompanying paraphyses. The spores are four-celled, dark brown, deeply constricted at the septum and fragment into four one-celled spores at maturity. Each cell has a diagonal germ slit. Kowalski (**1965a**) found a similar development in *P. typharum*. These characters are similar to those of the genus *Sporormia* which Müller and von Arx (**1962**) place in the family Sporormiaceae in the order Pseudosphaeriales.

Preussia multispora has rather light colored ascospores which are thirty-two in an ascus. It was described as *Anixiopsis multispora* by Saito and Minoura and transferred to the genus *Pseudeurotium* by Stolk (**1955**). Cain (**1961a**) considered the ascospores as four-celled and fragmenting early into thirty-two one-celled spores, and placed the fungus in the genus *Preussia*.

Synaptospora: Cain (**1956c**) described this genus as having black carbonaceous roughened cleistothecia. The asci are long cylindric and arranged in a parallel manner within a hymenium. The fruit body may have a minute pore-like ostiole. The spores are one-celled and brown but become united into chains of two, four, and eight cells. The asci have a thickened apical wall with the canal surrounded by thickening as in the Sordariaceae. The genus shows relationship to both *Preussia* and *Pycnidiophora*.

ELAPHOMYCETACEAE:

The genera *Elaphomyces* and *Mesophellia* are underground (hypogeous) fungi. In *Elaphomyces*, the ascocarps are globose to irregular in shape and have a complex outer wall or cortex usually of three layers. The outer layer may be yellow to brown or black and smooth, granular or echinulate. The inner core is cottony and consists of sterile dissepiments separating chambers filled with asci which are evanescent and fill the cavities with spores. The spores are globose and dark colored. Species of *Elaphomyces* are parasitized by several species of *Cordyceps. C. ophioglossoides* and *agariciformia. Mesophellia* differs in having a leathery or woody core and ellipsoidal, hyaline or light colored spores. The genus is confined to Australia.

TRICHOCOMACEAE:

This is a monotypic family with only one species, and is found in warmer regions. *Trichocoma paradoxa* has a fruiting body with a sterile woody stipe-like base and a saucer-like apex. The asci are borne in tubular cavities, separated by a thin peridium, within this cup-like apex. The asci are borne at the base of the tubular cavities, and the spores are pushed upward and outward.

MELANOSPORALES

The Melanosporales are intermediate between the Plectascales and the Hypocreales or Sphaeriales. In fact some authors have placed the families included here in the Plectascales. The perithecia have ostioles which are often elongate, but the arrangement of the asci in the perithecial centrum varies from stellate central clusters as in some Plectascales to simple hymenial arrangements about the periphery of the central cavity. The asci are evanescent and the ascospores are freed within the perithecial cavity and are pushed out through the ostioles as a spore mass or spore horn.

Three families are recognized.

I. Perithecia covered with spiral or branched hairs or setae .... CHAETOMIACEAE
II. Perithecia smooth or with simple hairs or setae, mostly about the ostiolar region
  A. Spores pale straw color to red brown .... MELANOSPORACEAE
  B. Spores hyaline.................................... CERATOCYSTACEAE

MELANOSPORACEAE:

This family has perithecia with more or less elongate ostioles and setae or tomentum often about the ostioles. The ascospores are one-celled, pale straw color to dark brown. The asci are evanescent and the spores are pushed out as a slimy mass. Perithecial development shows an evolutionary series of forms from those with a central radiate fascicle of asci to those with a simple or well developed hymenium.

I. Spores pale straw color mostly 3-8 × 1.5-5 μ; conidial state *Scopulariopsis* .... MICROASCUS
II. Spores dark brown mostly more than 8 μ long
  A. Spores 7-11 × 4-6 μ imperfect states *Graphium* and *Sporotrichum* .. PETRIELLA
  B. Spores mostly 10-30 μ long (some isodiametric and smaller); conidial states as bulbils or phialids. .... MELANOSPORA

Microascus: The species of this genus are found on soil dung, seeds and vegetable remains. The perithecia often have elongate ostioles and tomentum or setae about the ostioles. The spores are straw color and may be ellipsoid, bean-shaped or triangular, and smaller than in *Petriella*. The perithecia originate as a coiled several-celled archicarp around which vegetative hyphae form a stromatic wall. The ascogenous system arises from ascogenous cells of the archicarp and comes to lie in a central cavity. Sterile hyphae grow radially from the inner wall cells as a primitive type of paraphyses. The asci are formed as a radiate cluster which grow out between these paraphyses (Doguet, **1957a**; Corlett **1963, 1966**). The conidia are borne in chains as in the form genus *Scopulariopsis*, (Emmons and Dodge, **1931**) or on phialides from vegetative hyphae.

Microascus stysanophorus is intermediate between this genus and *Petriella*. It has large spores (9-10 × 5-6 μ) but they are pale straw in color. The conidial states are of the form genera *Stysanus* and *Sporotrichum*.

Microascus cinereus (*M. linasporus*?) has been isolated from infections of the human hand and from chromoblastomycosis infections in association with *Hormodendrum pedrosi*.

This genus and *Petriella* have both been monographed by Barron, Cain and Gilman (**1961, 1961b**).

Petriella: This genus is similar to *Microascus* but differs in the dark red-brown color of the spores. The ostioles are often elongate and have tomentum or hairs about their surface. The associated conidial states are *Graphium* (*Sporocybe*) and *Sporotrichum*.

Melanospora: In *Melanospora* the perithecia arise as a coil of several cells which are multinucleate (Doguet, **1955a**), except in *M. tiffanii* in which the archicarp is an elongate, tapered, many-celled gametic hypha (Kowalski, **1965**). The enveloping hyphae form a solid parenchymatic centrum which is differentiated to an outer perithecial wall and what Doguet terms a sous-hymenial (lower) and sus-hymenial (upper) thinner-walled pseudoparenchyma. The ascogenous system develops in a cavity between these two parenchyma tissues. Ingrowing paraphyses are not as well developed as in *Microascus*, the asci dissolving their way through the sus-hymenium. In *Melanospora fallax, M. moreaui,* etc., the asci form a radiate cluster from a stalked base of the sous-hymenium. In *M. zonata* the asci form a flat hymenium at the base of the centrum and in *M. parasitica* the hymenium is peripheral around the base and sides of the centrum as in the higher Pyrenomycetes. The spores are dark red-brown and mostly large ellipsoid (10-30 μ long) but some are smaller, cuboid, cylindric or larger with reticulate ridges on their walls. They are freed early from the asci and extruded in a slimy mass.

Many species of *Melanospora* show the phenomenon of commensalism. Such species will not grow or will not form perithecia when grown alone, but must be grown in association with certain other fungi which produce some stimulating substance. *Fusarium avenaceum, F. orthoceras, F. poae, Aspergillus amstelodami, Papularia sphaerosperma* and many other fungi produce stimulatory substances. Extracts of cultures of these fungi also stimulate the formation of perithecia when added to *Melanospora* cultures.

CHAETOMIACEAE:

This family is based upon the ornamentation occurring on the perithecia. The tomentum consists of irregularly coiled or tightly spiraled hairs or of stiff simple or branched setae. The family shows relationship to the Melanosporaceae in the evanescent walls of the asci which release the colored spores into the perithecial cavity, and to the Ceratocystaceae in the elongate ostioles of some genera. These fungi are capable of breaking down cellulose and are commonly found on such substrata as seeds, straw, paper and dung. They may also cause a deterioration of cotton fabrics. *Scopulariopsis* and *Verticillium* conidia have been reported on the hairs of some species and a *Scopulariopsis* state has been reported by Corlett (**1965**) for *Chaetomium trigonosporum.*

I. Perithecia with short stout ostioles
   A. Conidia occurring on the perithecial setae .... ASCOTRICHA
   B. No such conidia on the perithecial setae...................... CHAETOMIUM
II. Perithecia with elongate ostioles
   A. Perithecia covered with fine hairs .... LOPHOTRICHUS
   B. Perithecia with stout stiff spines .... CHAETOCERATOSTOMA

Chaetomium: This genus contains the most abundant species of the family. They are found as contaminants of seeds, may cause a deterioration of cotton fabrics and as a contaminant of laboratory cultures, particularly on filter paper.

The perithecium arises as a spiral archicarp around which enveloping hyphae from the base form the wall. The ascogenous hyphae lie in the base of the perithecial cavity and delicate sterile hyphae grow in from the sides or top of the inner wall cells, but these deliquesce as the asci mature. The asci are clavate and have a thin wall which breaks down and releases the spores into the perithecial cavity. The ascospores are pale olive to dark brown, ellipsoid, limoniform or cuboid and are pushed out through the short stout ostiole as a spore mass which is held in the hairs or setae about the ostiole.

The genus has been monographed by Chivers (**1915**) and Ames (**1961**).

Ascotricha: This genus was segregated by Chivers (**1915**) on the basis of the spherical conidia found on the perithecial hairs. The asci are cylindric. It has not been recognized by all later authors.

Lophotrichus: This genus differs from *Chaetomium* in the elongate ostioles with penicillate hyphae at the apex, in which characters it resembles *Ceratocystis*. The perithecia are covered with a hair-like tomentum.

Chaetosceratostoma: This genus also has elongate ostioles, approaching *Ceratocystis* in this character. Its perithecia are covered by stiff spines. According to Doguet (**1955**) the perithecium arises as an upright branch which becomes curved and spirals back downward around itself. It becomes septate into one- or two-nucleate cells. Sheathing hyphae arise from the three to four basal cells. They envelop the archicarp and form a bulbous base and an upward growing sheath which elongates to form the ostiole with a penicillate apex. The ascogenous hyphae develop free in the base of the perithecial cavity, not in a stromatic centrum as in *Melanospora*. The original archicarp disappears. The asci are formed from croziers in a basal fascicle. The ascus wall breaks down and frees the brown, lemon-shaped, flattened spores which have a germ pore at one end.

CERATOCYSTACEAE:

The family Ceratocystaceae is of considerable economic importance. Several species are the cause of diseases of crop plants, and others cause diseases of forest and shade trees. A number of species are associated with discoloration of the wood of standing trees and of cut lumber.

The perithecia are mostly dark in color and have elongate ostioles with a penicillate apex. The ascus walls break down easily and the hyaline spores are released into the perithecial cavity.

I. Perithecia dark colored; spores hyaline .... CERATOCYSTIS
II. Perithecia light colored, yellow to orange; spores light colored . SPHAERONAEMELLA

Ceratocystis: This genus is found in the earlier literature under *Ceratostomella* or *Ophiostoma.* Because of the evanescent character of the asci and the collection of the spores in a mass at the apex of the ostioles, some species of this genus were at first described in the Fungi Imperfecti under the form genus *Sphaeronaema.* A number of conidial states have been connected with species of *Ceratocystis* and several genera have been segregated on this basis. Those species which form endoconidia from phialides have been placed in a genus *Endoconidiophora.* Those species with coremium-like conidial states in *Leptographium* have been placed in the genus *Grossmania* (Goidanich, **1936**). Other species have a *Graphium* or a *Cladosporium* or *Cephalosporium* conidial state.

Several writers studied perithecial formation in *Ceratocystis,* and have found that ascus formation is unique. Elliott (**1925**) claimed that in *C. fimbriata* there was a stalked, one-celled "oogonium" with a trichogyne which fused with an antheridium followed by nuclear fusion. The asci arose as binucleate swellings which separated from the ascogenous hyphae after nuclear fusion. Mittmann (**1932**), Andrus (**1936**), Andrus and Harter (**1933**) and Gwynne-Vaughan and Broadhead (**1936**) in their studies of *C. fimbriata* found that the perithecium arose as a side branch which became coiled and divided into uninucleate cells. No antheridium nor any definite nuclear fusion was seen. Several cells of the ascogenous hyphae enlarge and become multinucleate. Lateral branches from the lower cells of the archicarp envelop it to form the perithecial wall. The large central cells of the archicarp become lobed and fragment as nuclear division proceeds. These cells come to lie free in a nutrient medium in the perithecial cavity. Their nuclei increase in number and the cells divide until ultimately a number of binucleate cells result. These cells develop directly into asci while lying free in the perithecial cavity. The only wall possessed by these asci is the original cell membrane. The asci develop within the vacuole of these cells and the delicate ascus wall breaks down, releasing the hyaline spores which are pushed up through the ostiole and collect in a spore mass in the penicillate hairs at the apex of the ostiole.

Such species as *C. piceaperda, C. ips, C. piceae, C. cana, C. pilifera* and others are the cause of blue stain and other discolorations of trees and lumber, particularly of coniferous trees. These species are associated with bark beetles. The mycelium grows in the sapwood and cambial layers and causes a discoloration, which does not, however, greatly affect the mechanical strength of the wood, but makes it unsuitable for furniture and cabinet work. The mycelium grows in the galleries of the bark beetles and the conidia are picked up by the beetles and distributed to healthy trees. In some cases the beetles feed upon the conidial coremia in the galleries.

Ceratocystis (*Ceratostomella*) *ulmi* is the most important species on hardwood trees. It is the cause of the so-called "Dutch Elm Disease". The disease was first studied in Holland and was known only in its conidial state, *Graphium ulmi,* which is the usual fruiting state on the host. Buisman (**1929, 1932**) obtained the perithecial state by crossing strains in culture, thereby proving it to be heterothallic. The mycelium grows in the sapwood and plugs the sapwood with a black gummy substance. In cross sections of twigs or stems this deposit appears as a ring of black-brown spots in the sapwood. This clogging of the vessels causes a wilting and drying of the leaves on infected branches. As this proceeds for several years into the larger limbs and trunk the trees eventually die. This disease has practically wiped out the elms in Europe and has caused great damage in eastern North America and is advancing into the midwestern states where many elms have been killed. Attempts to develop resistant strains have not been sucessful. The Asiatic elms, however, are immune. The disease is distributed by bark beetles of the genus *Scolytes*. Spraying by DDT has been used to control the beetles, but for ultimate prevention, all dead or infected trees should be cut and burned, for the beetles live in the dead bark.

Ceratocystis (*Endoconidiophora*) *fagacearum* is the cause of the "Oak Wilt Disease" which causes a great deal of damage to city plantings of oak and also attacks *Castanea* and *Castaneopsis*. The conidial state is *Chalara quercina*. The mycelium grows in the sapwood as in *C. ulmi* and causes a similar wilting and killing of leaves on infected branches. As the trees die mats of mycelium are formed beneath the bark and break through the periderm. Numerous endoconidia are formed on these mats and are carried about by insects. The species is heterothallic and if plus conidia are carried to minus mats (or vice versa) perithecia are formed.

Ceratocystis (*Endoconidiophora*) *fimbriata* forma *platani* is the cause of a canker stain of sycamore trees. It spreads slowly but eventually girdles the limbs and kills the tree.

Ceratocystis fimbriata is the cause of black rot of sweet potatoes, and is found in the tropics as a moldy rot of rubber, on coffee, etc.

Ceratocystis adiposa causes a disease of seed pieces of sugar cane and other tropical crops.

Ceratocystis paradoxa is also widespread in the tropics as a leaf rot of coconuts and as rots of sugar cane and pineapples. The above four species of *Ceratocystis* have conidial states in the form genus *Thielaviopsis*.

Heterothallism has been reported in a number of species of *Ceratocystis*, but irregularities occur, giving rise to both self-fertile and self-sterile strains.

The genus *Ceratocystis* has been monographed by Hunt (**1956**).

Sphaeronaemella: Sphaeronaemella Karsten was described in the Fungi Imperfecti, because the early deliquescence of asci freed the spores which were misinterpreted as conidia. Seeler (**1943**) saw asci and transferred the genus to the Ascomycetes near *Melanospora*. The perithecia are light colored, yellow to orange-brown, with an elongate penicillate ostiole. The asci are evanescent and the ascospores are allantoid and hyaline. The type species, *S. helvellae,* occurs on *Helvella infula. Sphaeronaemella fimicola,* which was studied by Cain and Weresub (**1957**), occurs on dung. They found that the formation of perithecia was stimulated by the presence of *Aspergillus repens* or of extracts of cultures of this species. The perithecia originate as an archicarp of three to five cells with an elongate trichogyne which fuses with vegetative hyphae either apically or laterally. Long chains of fusoid conidia are formed in culture. Von Arx and Müller (**1954**) placed *Sphaeronaemella* as a synonym of *Melanospora.* Hunt (**1956**) considered the genus similar to *Ceratocystis* but not identical to it. Cain and Weresub also considered it similar to but not congeneric with *Ceratocystis.* This is another genus which shows characteristics of both the Melanosporaceae and the Ceratocystaceae.

XYLARIALES

The family Xylariaceae originally included only stromatic forms of the Sphaeriales (Ascohymeniales) with numerous perithecia imbedded in a well developed stroma. Recent studies have indicated that a number of genera from several families, with perithecia borne singly, are closely related to the stromatic Xylariaceae. These families are here united under one order, the Xylariales.

There are so many variations in spore form and perithecial arrangement, and so many intermediate forms that it is difficult to sharply differentiate these families. The following key attempts to evaluate these differences.

I. Perithecia more than 1, with a differentiated wall, immersed in a well developed stroma. Mostly saprophytes on dead or decaying wood (a few on dung or as parasites on higher plants)... XYLARIACEAE

II. Perithecia borne singly, superficial or immersed in substrate or beneath a clypeus or poorly developed stroma. Mostly saprophytes on dung or vegetable remains.

A. Spores vermiform, hyaline, sometimes septate, rarely colored, perithecia mostly tomentose or setose. On dead vegetable remains......... LASIOSPHAERIACEAE

B. Spores ellipsoid to discoid, brown, with germ pores or slits

1. Spores with gelatinous envelopes or primary cell-like appendages or secondary gelatinous appendages. Mostly on dung, a few on vegetable remains..... .... SORDARIACEAE

2. Spores without appendages or envelopes. Mostly on dead wood or vegetable remains.. ROSELLINIACEAE

The common denominator for all these genera is the thin-walled ascus which may have an apical thickening of the wall with a central canal and sometimes a cap-like "apical apparatus." The spores are dark brown, one-celled with either germ pores or germ slits. The one exception to this is the family Lasiosphaeriaceae in which the spores are vermiform and hyaline. These spores, however, are of the same form and color as the immature spores of some species of *Podospora*, to which they seem related. They do not, however, have the ellipsoid swollen apical cell of most *Podospora* species.

The separation of the Roselliniaceae and the Sordariaceae is admittedly difficult and more or less artificial. There are species occurring on herbaceous stems and woody substrata which have the gelatinous appendages of the Sordariaceae but are closely related to such genera as *Rosellinia* and *Anthostoma.*

Paraphyses are usually present in this order, although they may be evanescent in the simpler forms.

SORDARIACEAE:

This family was originally erected for a group of fungi occurring on dung. There are, however, many other unrelated fungi which occur on dung, and there are other species occurring on vegetable debris which are closely related to the dung inhabiting forms. Cain (**1934**) has pointed out that most species of *Delitschia*, formerly placed in the Sordariaceae, have an ascus with a double membrane and a bitunicate method of spore ejection. Munk (**1957**) and von Arx and Müller (**1954**) and Müller and von Arx (**1962**) have placed several such genera with septate spores in a family Sporormiaceae in the Ascoloculares. There has been a great deal of confusion as to the use of generic names within the family. As circumscribed here, the family includes those genera which have spores with a gelatinous envelope or "primary cell-like appendage or "secondary" gelatinous appendages. The following key is adapted from those of Cain (**1934, 1962**), von Arx and Müller (**1954**), and Müller and von Arx (**1962**).

I. Spores with a single dark brown cell
  A. Perithecia in a stroma or beneath a clypeus
    1. Spores with a gelatinous envelope. . . . . HYPOCOPRA
    2. Spores with appendages . . . . ENTOSORDARIA
  B. Perithecia without a stroma
    1. Spores with a single ellipsoid brown cell
      a. Spores with a gelatinous envelope or secondary appendages (no primary appendage) . . . . SORDARIA
      b. Spores with a primary, cell-like, appendage
        i. Perithecia cylindric, thick-walled; spores vermiform at first with secondary appendages, when mature with an upper ellipsoid brown cell . . . . BOMBARDIA
        ii. Perithecia pyriform leathery; spores sometimes vermiform when first delimited with an upper ellipsoid cell which turns brown at maturity . . . . PODOSPORA
        iii. Spores triangular with a short primary cell-like appendage. . . . . TRIANGULARIA
II. Spores 2-celled with secondary appendages . . . . ZYGOSPERMUM

The coprophilous species have membranous-walled perithecia with short ostioles. The spores are shot out with some force and the gelatinous appendages aid in the adherence of the spores to neighboring grasses and herbs and are then picked up by grazing animals and pass through their intestinal tracts and germinate on the dung. The perithecia often have a fine tomentum. The asci are thin-walled and either eight- or many-spored. Members of the genera *Sordaria*, *Podospora* and *Bombardia* have been used extensively for the study of genetic behavior.

The family has been monographed by Griffiths (**1901**) and Cain (**1934**).

Sordaria: This genus is common on the dung of grazing animals, such as cow, horse, sheep, deer, moose, etc. It is easily obtained in dung cultures in the laboratory. The perithecia are superficial or slightly immersed and often covered with a fine tomentum, which may disappear with age. The ascospores are ellipsoid, one-celled, brown and have a gelatinous sheath or secondary appendages, which are gelatinous, at one or both ends of the spore. Several species have polysporous asci. These have been placed in a separate genus, *Philocopra*, by some. Several species of *Sordaria* have been reported as contaminants of seeds (Cain, **1948a**).

Sordaria fimicola and *S. macrospora* have been studied extensively as to their genetic behavior. These two species are homothallic, self-fertile and have eight haploid ascospores in their asci. Ascogonia are produced but no antheridia or conidia occur. The nuclei divide conjugately in the ascogonium and migrate into the ascogenous hyphae. Crosses can occur between different mycelia. A certain number of nuclei migrate from one mycelium to the other at the point of contact. The hybrid perithecia which occur are identical to those of the ascogonial parent. They can be detected only by the presence of marker genes, such as those determining color in the ascospores. In hybrid perithecia these marker genes are segregated and occur as ascospores of different colors in the asci. Some fifteen genes have been detected which interfere with the morphological development of the various stages of the perithecium. All but one of these result in the formation of sterile fruit bodies. Sterility also occurs when certain geographic races are crossed with one another.

Podospora: The perithecia of this genus are also abundant on the dung of grazing animals. They may be tomentose or have a crown of setae about the ostiole. The spores, when first cut out in the ascus, are cylindric, and vermiform in a few species, but in most species consist of an ellipsoid upper portion and an elongate cylindric lower portion which is finally cut off by a septum. The wall of the upper cell becomes thickened and dark brown. The lower cell remains as a hyaline cylindric "primary" appendage. Gelatinous secondary appendages may also occur, attached to the lower end of the primary appendage, the upper end of the ellipsoid cell or variously arranged about the septum between the two cells. In *Podospora zygospora* there are two ellipsoid brown cells connected by a spiral, hyaline, primary appendage. The genus has been emended and new combinations made by Cain (**1962**).

Podospora anserina and *Bombardia* (*Podospora*) *lunata* have also been used in genetical studies. *Podospora anserina* has four ascospores, which are heterokaryotic, in the asci. Most such spores contain both plus and minus nuclei and give rise to self-fertile mycelia. Occasional uninucleate ascospores occur which give rise to heterothallic mycelia, as in *Neurospora*. Ascogonia and pear-shaped spermatia in spermagonia are formed on all mycelia. No macroconidia occur. The ascogonia are uninucleate; the spermatia fuse with the trichogyne and their nuclei migrate into the ascogonia. The uninucleate ascospores give rise to one of two heterothallic strains (plus and minus). Crosses between races from such uninucleate ascospores have shown a curious behavior known as "non-reciprocal" and "reciprocal" incompatability. This has been shown to be due to the superimposition of two pairs of allelic genes upon the ordinary bipolar heterothallic ones. In the non-reciprocal cross, one race (minus) has the genotype a, b and the other race (plus) the genotype a_1, b_1. The spermatia of the a_1b_1 race can fertilize the a, b ascogonia, but the a, b spermatia cannot fertilize the a_1b_1 ascogonia (i.e. the reciprocal cross). If the trichogyne of an a, b ascognoium grows toward an a_1b_1 spermatium, it will fuse with it and the trichogyne nuclei degenerate. However, if the trichogyne grows toward a spermatium with the b gene (a, b), the b gene prevents fusion with the trichogyne, the nuclei do not degenerate, and the cross is sterile for the trichogyne grows on past the spermatium. In the case of reciprocal incompatability, two additional pairs of allelic genes are present (i.e., a, b, c_1, v_1 and a_1, b_1, c, v), in which case both the b and the v gene prevent spermatial fusion with the trichogyne and all crosses are sterile.

In *Bombardia lunata* there are eight haploid ascospores in each ascus, and the resulting mycelia are heterothallic (plus or minus) and monoecious, that is, they have both ascogonia and spermagonia. The plus and minus spermatia release substances which attract the trichogyne of the ascogonium. If a plus trichogyne meets a minus spermatium, fusion occurs, but if it touches a spermatium of the same strain (plus) there is no fusion between the trichogyne and the spermatium. In other words, the incompatible gene prevents fusion, mere contact is not sufficient.

Bombardia: The genus *Bombardia* is intermediate between *Podospora* and *Lasiosphaeria* and the distribution of species depends upon the authors. The perithecia occur on both dung and decaying vegetable remains. The type species has elongate cylindric perithecia with very thick walls, looked upon by some as a stroma. The asci are perforated at the apex with a refractive globule near the upper end. The spores are elongate cylindric, bent at the lower end and with a hyaline secondary appendage at each end. At maturity the upper end swells into an ellipsoid cell which may become brown as in *Podospora*. The species with leathery thin-walled perithecia, in contrast to the thick-walled carbonaceous perithecia of the type species *B. fasciculata*, have been placed by Chenantais (**1920**) and others in the genus *Lasiosordaria*, which is very similar to *Podospora*.

Triangularia: In this genus the ascospores are at first more or less spherical, but finally become three-sided forming a pyramidal shape. A septum is laid down across the base of the pyramid, this lower cell remains hyaline as a primary appendage and the upper cell becomes dark brown. The one species *T. bambusae* is found on bamboo.

Zygospermum: In this genus, which occurs on dung, the spores consist of two dark brown ellipsoid cells with a germ pore at each end and a long gelatinous secondary appendage at each end of the spore.

Hypocopra: The perithecia of the species of this genus are immersed in the dung and usually have a filamentous or well developed dark colored stroma in which they are imbedded. The spores are one-celled, ellipsoid, dark brown with a longitudinal slit and a gelatinous envelope.

Entosordaria: The perithecia of this genus are found on woody stems and are imbedded beneath a clypeus of modified host tissue, in which respect it is similar to *Anthostoma*. The spores, however, have a small hyaline primary appendage below the upper brown portions of the spore.

ROSELLINIACEAE:

In the Roselliniaceae as outlined here there are included a number of genera which various mycologists have considered to be related to the Xylariaceae in the character of the spores and the asci. They have previously been placed in various families (Melanommeae, Sordariaceae, Nectriaceae, etc.) and show some overlap with the Roselliniaceae and the Sordariaceae, in morphology and habitat. The perithecia are globose to pyriform, superficial or immersed, but occur singly. The spores are one-celled brown with germ pores or germ slits, but without the gelatinous appendages (except in *Cainia*) or envelopes. They are found on vegetable debris, woody substrata or on dung. There are numerous paraphyses about the asci in these genera.

I. Perithecia borne superficially on the substrate
  A. Ascospores with various wall markings
    1. With vertical striae on the ascospores .... NEUROSPORA
    2. With circular pits upon the ascospores .... GELASINOSPORA
    3. With coma-like striae on the ascospores................ NEOCOSMOSPORA
  B. Ascospores without such markings
    1. Perithedia glabrous, ascospores lacking pores or slits .... ROSELLINIA
    2. Perithecia often with setae or tomentum; ascospores with longitudinal germ slits .... CONIOCHAETA
II. Perithecia immersed in the substrate
  A. Ascospores with 4 equatorial germ pores .... AMPHISPHAERELLA
  B. Ascospores without equatorial germ pores
    1. Parasitic upon *Clavaria* .... HELMINTHOSPHAERIA
    2. Perithecia saproplytic, on woody or herbaceous stems........ ANTHOSTOMA

Rosellinia: This genus was originally palced in the Melanommeae and a number of subgenera were described, several of which have since been raised to generic rank. They occur mostly on decaying wood substrata. The name *Rosellinia* is accepted in the sense of von Arx and Müller (**1954**) for those species with large perithecia, over 500 μ in diameter. These often occur upon a subiculum or with a thick double wall, the outer layer of which is stromatic in character (i.e. *R. aquila*). They may also be closely crowded and grade off into the genus *Hypoxylon*. *Rosellinia necatrix* is found on the roots of grape and may do some damage to this crop.

Coniochaeta: This was originally a subgenus of *Rosellinia* and was raised to generic rank by Traverso (**1907**). It differs in the smaller perithecia (under 300 μ in diameter) and in the presence of spines or hairs on the perithecia in most species. They occur on dung or on decaying vegetable remains and grade off into the Sordariaceae. The spores are more or less discoid and have a longitudinal germ slit along the narrow edge. There are numerous paraphyses about the asci.

Amphisphaerella: Amphisphaerella was originally placed as a subgenus of *Rosellinia* because the perithecia are more or less sunken in the substratum. Kirschstein (**1933**) raised it to generic rank. Munk (**1953, 1957**) used the presence of four germ pores in the equatorial region of the ascospores as a generic character. The two species known are found on woody substrata.

Helminthosphaeria: This genus consists of a single species parasitic upon species of *Clavaria*. The perithecia are partially immersed in the host tissue and surrounded by a stromatic development of dark brown hyphae, and have setose hyphae about the ostiole. Emergent conidiophore hyphae cut off two-celled brown conidia. The spores have a longitudinal germ slit.

Anthostoma: The genus *Anthostoma* includes forms on woody or herbaceous stems in which the perithecia are immersed singly or in groups, but with separately erumpent ostioles, which may be somewhat elongate. As outlined by von Arx and Müller (**1954**) it includes some ten other genera as synonyms.

Neurospora: The genus *Neurospora* has become very important to the geneticist and the biochemist. The perithecia were unknown until Shear and B.O. Dodge (**1927**) crossed certain conidial strains and obtained them in culture. The perithecia are pyriform, superficial, dark blue black and were first placed in the Hypocreales because of their color. In *Neurospora sitophila* and *N. crassa* there are eight spores in the ascus which are uninucleate when cut out but undergo mitosis to form a binucleate spore. These spores give rise to two strains (a and A) of heterothallic mycelia (Dodge, **1927**). *N. tetrasperma* has asci with four "giant" spores which are binucleate when cut out and four-nucleate at maturity and give rise to a homothallic mycelium which contains both the "a" and "A" type of nuclei. Such mycelia can form perithecia without fertilization by a compatible nucleus (Dodge, **1928**). Occasionally "dwarf" spores are formed with less than four nuclei, in an ascus with more than four spores.

The genus has two types of conidia. The macroconidia, *Monilia sitophila*, are large, ellipsoid conidia which are borne in branching chains from clusters of branching conidiophores. They are bright orange in color and cause a troublesome "Bakers' mold" for they are heat resistant to a temperature of 130 F and may live over in the center of a loaf of bread during baking. The microconidia are much smaller, globose to ellipsoid and formed endogenously from small phialids on much branched conidiophores. They have a large nucleus and a small amount of cytoplasm and do not germinate readily.

The perithecia of *Neurospora* arise as a short terminal coil of four to five cells. Hyphae from the base grow up around this coil to form a tight ball of hyphae with a number of filiform hyphae growing out from it, which have been termed trichogynes. This "protoperithecium," as it is called, may reach some size but will never mature to form asci or ascospores until fertilized by nuclei of the opposite compatible strain.

If microconidia of strain "a" are placed upon mycelia of strain "A" containing protoperithecia (or vice versa), they will fuse with the hyphae and nuclei will pass into the mycelia and reach the ascogonium. These perithecia will then proceed to mature and form asci and ascospores. Similarly, if macroconidia are placed on mycelia of the opposite strain, the protoperithecia will mature. If strain "a" is placed on one side of a petri dish and strain "A" on the opposite side, when they grow together perithecia will mature on one strain first, and a few on the opposite strain later. Dowding and Gowan (**1942**) and B.O. Dodge (**1935**) have shown that the hyphae of one strain will fuse with those of the opposite strain, and nuclei of one strain will migrate into the hyphae of the opposite strain and eventually reach the ascogonium of a protoperithecium, which will then mature. This migration is usually in one direction at first.

The ascospores of *Neurospora* are ellipsoid, one-celled and brown, with fine parallel striae on the wall, and lie in a single row in the ascus. B.O. Dodge developed a technique of isolating single ascospores in the order in which they lie in the ascus. By numbering the spores in sequence as they lie in the ascus (1-2-3----8) and growing the mycelium from single spores, the segregation of the gene characters of the two parents can be determined. Dodge (**1928a**) was able to obtain recognizable mutants of *N. sitophila* and *N. crassa* and was able to hybridize not only these mutations but also the species of *Neurospora* (Dodge, (**1931**).

He pointed out that there are three possible arrangements of pairs of allelomorphic genes within the ascus. If reduction division occurs in the first division of the diploid nucleus the distribution of the two allelomorphs will be a four-four arrangement (four of one allelomorph in one end of the ascus and four for the other in the other end). A two-two-two-two arrangement indicates a reduction division in the second division.

One of the most striking of Dodge's (**1930**) mutants was an albino strain which produced no macroconidia. The colonies were therefore pure white in contrast to the bright orange of the wild strain. When a wild strain "a" was crossed with an albino strain "A", each ascus in the hybrid perithecia produced four ascospores which gave albino mycelium and four which gave orange mycelium, of the strain. In some cases he found that the albino and wild strain spores were arranged in a four-four position, but that the sex characters (a and A) were in two-two-two-two arrangement. This indicated to him that there were two reduction divisions in the ascus and supported the theory of brachymeiosis of Gwynne-Vaughan and Harper which proposed two fusions, one in the ascogonium and one in the young ascus, giving a tetraploid nucleus.

Dodge studied the cytology of the ascus in *N. tetrasperma* and showed that the nuclei were normally arranged in such a fashion that each spore was cut out around two nuclei, one of type "a" and one type "A". Such "giant" spores gave rise to mycelia which contained nuclei of both sex types and could produce perithecia without receiving nuclei from a compatible strain. Such mycelia appear to be homothallic, but the species is what has been termed pseudohomothallic. This is because often "dwarf" spores with nuclei of only one sex type are cut out and then give rise to mycelia which are truly heterothallic.

Lindegren (**1932-1949**) in his studies of *N. crassa* obtained many mutants and analysed thousands of asci as to the arrangement of spores. He found, as did Dodge, that sometimes one set of allelomorphs showed a four-four arrangement whereas a second pair might show a two-two-two-two arrangement. He did not believe, however, that there were two reduction divisions. He pointed out that meiosis consisted of two nuclear divisions and that if there were a second reduction division it must occur in the third division. He never found a one-one-one-one----one alternate arrangement of allelomorphic genes which ruled out a second reduction division. He explained the four-four and the two-two-two-two arrangement of genes in the same ascus as being due to a cross-over in the four strand stage of the meiotic division. This theory of a single fusion of the nuclei of the ascus initial and the crossing over in the four strand stage has been generally accepted. Moreover the occurrence of a two-two-two--two arrangement indicates such a cross-over. Calculation of the percentage of such cross-overs between several allelomorphic genes enables the geneticist to determine the relative position of such genes on the chromosome. McClintock (**1945**) has found the number of chromosomes to be seven in *Neurospora*. A large number of mutant genes have now been plotted or mapped on the chromosomes of *Neurospora*.

The recent genetic work of Beadle and Tatum (**1941, 1945**), Houlihan, et al (**1949**) and many others, based in turn on the physiologic work of Schopfer (**1954**), Robbins (**1950**) and their students, on the vitamin needs of various fungi, has opened a most fascinating field of study which promises to clarify many of the basic mechanisms of living organisms. Schopfer (**1954**) was the first to show clearly that certain fungi could be grown on media composed of purified sugars and inorganic salts, whereas other strains could not. Such deficient strains needed the addition of certain organic compounds which were necessary to their metabolism but which they could not elaborate. He recognized three classes of metabolism in plants:

1. Phototrophic or green plants which could synthesize sugars from inorganic materials and also all other substances needed in their metabolism, from such sugars and inorganic salts.

2. Autotrophic plants without chlorophyll and therefore unable to synthesize sugars but able to produce all other organic substances needed in their metabolism.

3. Heterotrophic plants which not only lack the ability to form sugars but also lack the ability to synthesize certain other fundamental organic substances from sugars and inorganic salts.

Such heterotrophic plants (largely certain strains of fungi) must of necessity seek a substrate or host which will supply them the deficient substance. The wild strains of *Neurospora* are almost autotrophic, being deficient only in their ability to form the vitamin biotin, which must be present in the medium on which they grow.

Beadle in his early work was able to produce numerous mutations, both morphologic and physiologic, by exposing macroconidia of *Neurospora* to short wave radiation. Most of these mutations resulted in a loss of the ability of the strain to produce some necessary organic compound such as vitamins, amino acids, purines or pyrimidines. A "niacine-less" mutation, for instance, lost the ability to produce niacin and this substance in addition to biotin had to be added to media upon which the strain would grow. A large number of such deficiency mutations were obtained. It was found that the loss of the ability to form any one compound was due to the loss of a specific gene. On the other hand, in some cases a number of genes, variously situated on the chromosomes, were found to be necessary for the organism to synthesize certain organic compounds. This was due to the fact that some compounds are synthesized in a series of steps from precursors of the final compound. By using strains which were deficient for different genes affecting the various steps in the elaboration of various amino acids, proteins, etc., a fascinating field was opened up which indicated the various "pathways" by which organic compounds were built up in the metabolism of the organism.

A single example will suffice. Srb and Horowitz (**1944**) detected seven different strains which lacked a single gene concerned with the elaboration of arginine. Gene seven was in a strain lacking the ability to form arginine from citrulline. This strain must be provided with arginine before it could form protein. Genes five and six were in strains which lacked the ability to form citrulline from ornithine. Such strains can be grown by adding either arginine or citrulline. Genes one to four occurred in strains lacking the ability to synthesize ornithine. Such strains would grow if ornithine, citrulline or arginine were added (Srb, 1950).

This method has become an invaluable tool for the biochemist to trace the complex pathways in the elaboration of organic compounds in the metabolism of living organisms.

Gelasinospora: This genus is very similar to *Neurospora*, differing only in the occurrence of circular pits instead of vertical striae upon the ascospores. Most species are heterothallic, but *G. tetrasperma* is homothallic and has four-spored asci, and may produce dwarf heterothallic spores as in *Neurospora*. Spermatia have been described by Cain (1950) for *G. adjuncta*. The conidiophores consist of branching hyphae of short cells which produce minute terminal phialides from which small globose microconidia are extruded. The species occur on dung and as contaminants of seeds.

LASIOSPHAERIACEAE:

The single genus *Lasiosphaeria*, which is placed here, is related to the Sordariaceae. The perithecia are found on decaying wood or vegetable debris and bear various types of tomentum or setae. The asci are thin-walled with a thickened apex containing a refractive ring. The ascospores are similar to the immature ascospores of *Bombardia* and *Podospora*. They are cylindric, curved or "S" shaped and hyaline or sometimes pale brown. They are one-celled or septate and in some species have two terminal pointed appendages, but they do not have the ellipsoid dark colored cell found in *Bombardia* or *Podospora*.

XYLARIACEAE:

This family includes the stromatic forms with more than one perithecium in a stroma. The perithecia are separately erumpent (except in *Anthostoma*) and have a differentiated wall. The members of the family are found on decaying wood, roots and old stumps. A few may be weak parasites on weakened tissues. One genus is found on dung. They are widely distributed in both the temperate and tropical regions. The stroma is made up of mostly fungus tissue and has a differentiated surface crust or "ectostroma." The asci are cylindric, thin-walled and with a thickened apical wall through which runs a canal with a complex plug or "apical apparatus" which stains in potassium iodide. The ascospores are uniseriate, one-celled, dark brown and have germ pores or a germ slit. The species of Iowa have been monographed by Gilman and Tiffany (**1951**) and for Britain by Miller (**1930, 1932**).

I. Stromata crustose, effuse, pulvinate to globose or cylindric
  A. Internal tissues (entostroma) dark colored
    1. Internal tissues fibrous, in concentric layers; stroma globose to pyriform or short stalked .... DALDINIA
    2. Internal tissues not as above; perithecia monostichous
      a. Stromata concave, sunken .... NUMMULARIA
      b. Stroma not concave, effuse to globose
        i. Perithecia beneath a folded or plicate ectostroma .... CAMILLEA
        ii. Surface of stroma smooth, not plicate
          (1). Perithecia globose to ellipsoid, firmly seated in the stroma .... HYPOXYLON
          (2). Perithecia elongate, easily separable from the stroma and one another........... .... CAMAROPS
          (3). Perithecia single or grouped in subglobose erect extension of an effuse stromatic base .... KRETZSCHMARIA
  B. Internal tissues (entostroma) white, fleshy, or corky
    1. Stroma pulvinate to globose with a dark crust (ectostroma)......... PENZIGIA
    2. Stroma white throughout, without a dark ectostroma
      a. Spores with a sculptured epispore .... SARAWAKUS
      b. Spores without a sculptured epispore.................... SARCOXYLON
II. Stromata stiptate, erect, simple or lobed
  A. Stroma stiptate with a disc-shaped head bearing the perithecia, on dung. PORONIA
  B. Stromata erect, simple, lobed or branched, internally white fleshy....... XYLARIA

Hypoxylon: The genus *Hypoxylon* is a large and widely distributed one . The stromata are effuse, pulvinate or subglobose. They consist of a surface "ectostroma" which may be variously colored, white, brown, red, purplish or black, and an "entostroma" which is dark, coriaceous or carbonaceous.

The species of *Hypoxylon* are all found on woody substrata. The genus has been monographed by Miller (**1961**). He believed that many species may be weak parasites and fruit on the dead substrata. The mycelium proliferates between the bark and the periderm to form an ectostroma, which breaks through the overlying tissues to form a crust of often brighter colored stroma on the surface of which there is produced a layer of conidiophores bearing small ellipsoid to globose conidia. Other mycelia penetrate the bark and produce an entostroma which becomes erumpent and superficial, and within which the perithecia arise as coiled archicarps. The perithecia are separately erumpent by their ostioles, through the ectostroma.

Miller included *Ustulina* and most of the species of *Nummularia* in the *Hypoxylon*. He recognized four sections, as follows: 1. *Hypoxylon* with bright colored, leathery or woody stromata, and with the perithecia usually papillate prominent, and the ostioles umbilicate; 2. *Papillata* with white to gray or black carbonaceous stromata, prominent perithecia, but papillate ostioles; 3. *Annulata*, with prominent perithecia and ostioles surrounded by an annular disc, mostly tropical; 4. *Applanata*, stromata effuse, flat, carbonaceous, perithecia sunken, not prominent, ostioles papillate or umbilicate.

Hypoxylon pruinatum causes a bark canker of poplar (Povah, **1924**). Hubbes (**1962, 1964**) has shown that pyrocatechol and several other substances which inhibit the growth of *H. pruinatum* are found in the bark of species of poplar. The fungus does not grow on healthy bark but will grow in the sapwood. He concludes that the canker arises from injured or unhealthy bark through which the fungus reaches and grows in the sapwood. This is the characteristic habit of infection of many weak parasites.

There have been a number of genera segregated from *Hypoxylon* which are recognized by some authors and not by others.

Ustulina: U. vulgaris (*Hypoxylon deustum* accd. to Miller) is found on the roots and base of the stumps and logs of decaying wood. The stroma is indeterminate and follows the contours of the substratum. It is white and fleshy at first, bearing a layer of conidiophores and conidia. At maturity it becomes black and carbonaceous.

Nummularia: This name was originally used for species which arose beneath the periderm, which was thrown off to expose the stroma. Miller limits the genus to the type species, *N. discreta,* which has sunken, circular concave stroma. *Nummularia discreta* causes a very destructive blister canker of apple (and some other hosts) throughout apple growing areas (Cooper, **1917**). It is a wound parasite and penetrates into the wood, coming to the surface to form a sunken cankerous area. Discs of mycelium form beneath the periderm and produce numerous small conidia on their surfaces, which cause infection. Ascigerous stromata occur the following year, when the periderm is thrown off from these stromatic discs. Control is chiefly by prevention of insect and wound cracks. Cutting out of infected tissue is unsatisfactory for it is difficult to excise all active mycelium in the wood.

Daldinia: The genus *Daldinia* has black, globose superficial stromata, sometimes with a short basal stalk. The central entostroma is made up of concentric layers of a fibrous carbonaceous tissue. The stromata are found on dead and decaying wood. Ingold (**1954**) described the ascogenous hyphae of *Daldinia* as being up to forty μ long, unbranched and non-septate. The asci arise from croziers formed on short lateral branches.

Xylaria: In *Xylaria,* the stromata are upright, clavate, lobed or branched. They are commonly found growing from the roots or base of stumps and may cause a weak root rot. The young stromata are upright, white and fleshy and covered with a layer of conidiophores which bear hyaline conidia. The perithecia arise as coiled archicarps in the surface layers of this stroma. A blackened layer of ectostroma is formed over the surface of the stroma at maturity, the perithecia have differentiated walls and their ostioles are separately erumpent. Blackened strands of hyphae known as "rhizomorphs" may arise from infected roots and reinfect at some distance.

Penzigia: The stromata here are similar to those of *Hypoxylon*, but the internal entostromatic tissue is light colored and fleshy, as in *Xylaria*. The stromata are mostly hemispheric and attached centrally beneath by a short stalk.

Camillea: This genus has stromata which are also similar to an effuse *Hypoxylon* except that the surface is plicate or furrowed. The genus is not recognized by all authors.

Camarops Karst.: This has perithecia which are elongate cylindric and easily separable. Miller considers it distinct from *Bolinia* which von Arx and Müller (**1954**) consider a synonym.

Bolinia: Bolinia differs from *Camarops* in the semi-globose, polystichous perithecia. Both *Camarops* and *Bolinia* are found on dead wood and are mostly tropical in their distribution.

Kretzschmaria: In this genus the perithecia are borne singly or in small groups in erect stromatic outgrowths of a basal effuse stroma.

Poronia: This is an interesting genus occurring on dung. The stroma is stipitate with a disc-shaped head in which the perithecia are imbedded. The internal tissue is light colored and fleshy, as in *Xylaria*, and the spores have a gelatinous envelope. Its habitat and spores suggest a relationship to the Sordariaceae.

Sarcoxylon: Von Arx and Müller (**1954**) include a number of genera as synonyms of *Sarcoxylon*. They all have pulvinate to globose stromata which are white or light colored and fleshy both within and on the surface. Many of them have been placed in the Hypocreales because of their light color. The perithecia, however, have a dark differentiated wall and the spores are one-celled, brown and usually have a longitudinal germ slit as in the Xylariaceae.

Engleromyces goetsii, which is included here, forms globose white stromata as large as a man's fist, on the culms of bamboo.

The genus *Wawelia* consists of a single species *W. regia* only rarely found, on rabbit dung. It was placed in the Hypocreales by Namyslowski. Müller (**1954**) considered it a synonym of *Sarcoxylon*, but later transferred it (**1959**) to the Melanosporales. Doguet (**1961**) reported upon the development of the ascocarp and suggested its relation to *Melanospora*. This is an example of the difficulty of determining the relationships of some Ascomycetes. The light colored fleshy, cylindric stroma originally placed it in the Hypocreales. The fleshy stroma and the one-celled brown spores with germ slits relate it to the Xylariaceae. Furthermore the asci normally contain four ascospores, but may have two to six spores. This and its habitat suggest a relationship to *Neurospora* and the Sordariaceae. Doguet placed it in the Melanosporales because of the perithecial development, and the absence of any well developed "apical apparatus" in the ascus tip. The perithecia arise as a coiled archicarp in the upper portion of the stroma. The sterile tissue which develops about the archicarp consists of an outer wall layer of flattened cells and an inner matrix of thin-walled pseudoparenchyma. The ascogenous system finally comes to rest in a cavity within the base of this parenchyma. Paraphyses sprout out from the base and sides of the inner wall cells around this cavity and grow in radially with free ends forming true paraphyses. The asci grow up between these septate paraphyses. Doguet believed that the spores are freed early from the asci, within the cavity, as in *Melanospora*.

Sarawakus Lloyd: This genus differs from those placed under *Sarcoxylon* only in the sculptured network on the surface of the ascospores.

ALLANTOSPHAERIALES

Höhnel (**1918**) erected the family Allantosphaeriaceae with four sections, Diatrypeen, Valseen, Calosphaerieen and Coronophoreen. Fitzpatrick (**1923**) described the Nitschkieae as a sub-family, but could find no described family to which he believed it to be related. The Valseen have been placed as a family under the Diaporthales because of the type of ascus and the perithecial centrum. The remaining four families each show a divergent evolutionary trend, but have certain characters in common. The ascospores in most species are allantoid. The asci are clavate with a somewhat thickened apical wall and with a persistent, more or less elongate basal stalk holding the asci in a basal fascicle or persistent hymenium within the perithecial centrum, in contrast to the deliquescent stalks and loose arrangement of the asci of the Diaporthales. The species are all wood inhabiting. These families are placed here more or less arbitrarily in the same order, with a description of the divergent characters under each family.

I. Perithecia non-ostiolate; asci extruded in a mass by the swelling of a gelatinous tissue (Quellkörper) .... CORONOPHORACEAE
II. Perithecia not as above
  A. Perithecia superficial with a short stalk-like base .... NITSCHKIACEAE
  B. Perithecia immersed not stalked
    1. Perithecia clustered but not within a stroma .... CALOSPHAERIACEAE
    2. Perithecia imbedded within a stroma; often with a marginal blackened zone .... DIATRYPACEAE

DIATRYPACEAE:

In this family the perithecia are immersed in a stroma of various sorts within the bark of woody stems or on decorticated wood. The ostioles are disc-shaped to cylindric and usually sulcate with two or more deep grooves in the apex. The asci are clavate with a persistent stalk and form a definite hymenial layer at the base and sides of the perithecial centrum. The apical wall of the ascus is thickened and has a characteristic refractive ring which appears as two refractive spots in a lateral view. The ascospores are allantoid and pale yellow to brown, always appearing yellow brown in mass. The conidial states are of the form genera *Libertella*, with a wide open cavity or *Cytosporina* with an enclosed cavity. The conidia are of two types; one long filiform, curved to hamate and the other shorter allantoid, and hyaline. They are exuded in yellow to orange colored spore masses or spore horns (Wehmeyer, **1925**). Some species are weak parasites, attacking weakened or dying branches, but fruiting on the dead stems. A key to the genera follows.

I. Asci containing 8 spores
  A. Stroma effuse, ostioles separately erumpent
    1. Mostly on decorticated wood; surface of wood blackened; ascospores pale yellow .... EUTYPA
    2. On corticated wood, erumpent through the periderm; spores darker yellow
      a. Entostroma strongly developed, erumpent through the exfoliated periderm; ascospores 1-celled .... DIATRYPE
      b. Entostroma not erumpent; ostioles separately erumpent through the periderm; ascospores 1-5 septate, dark yellow-brown. . CRYPTOSPHAERIA .... (ENDOXYLA)
  B. Stromata disc-shaped, sharply margined; ostioles separately erumpent. DIATRYPE
  C. Stromata valsoid; perithecia grouped, ostioles collectively erumpent through a disc
    1. Ascospores 1-celled, allantoid, yellow-brown
      a. Perithecia few, 1-4, in a group, ostiolar disc not strongly erumpent .... QUATERNARIA
      b. Perithecia more numerous; disc more prominent............ EUTYPELLA

2. Ascospores yellow-brown to dark brown, 1-several celled
   a. Ascospores 1-celled, dark brown .... ANTHOSTOMA
   b. Ascospores 1 to several celled, dark yellow-brown, allantoid . ENDOXYLINA
   c. Ascospores 2-celled, dark brown, allantoid to ellipsoid .... VALSARIA

D. Perithecia erumpent on a thin basal stroma; ostioles elongate. Ascospores 2-celled brown .... RHYNCOSTOMA

E. Perithecia borne on an erect clavate to lobed stroma; ostioles short. Ascospores 2-celled and brown .... XYLOBOTRYUM

II. Asci with more than 8 spores
   A. Stromata pustulate-effuse, with many small groups of perithecia; disc not strongly erumpent .... CRYPTOVALSA
   B. Stromata isolated to pustulate, strongly erumpent; ostioles separately erumpent .... DIATRYPELLA

The genera *Endoxyla, Anthostoma,* and *Valsaria* include heterogeneous groups of species whose position in the grand scheme of the Euascomycetes is difficult to place. A careful study of these genera is needed. They are placed here in the Diatrypaceae with reservations.

Endoxylina: There has been a great deal of confusion concerning the genera *Endoxyla, Endoxylina, Eutypopsis*, etc., and von Arx and Müller (**1954**) used the name *Endoxyla* for species with elongate ostioles and one-celled ellipsoid ascospores and placed it in the Diaporthaceae, but some species described under this name have eutypoid stromata similar to *Cryptosphaeria* and have allantoid one- to two-septate spores. In 1962 these same authors (Müller and von Arx) placed *Endoxylina* in the Diatrypaceae and included species with allantoid, one-septate ascospores and scattered perithecia.

Endoxyla: Species of *Endoxyla* may have perithecia which are scattered singly in an effuse stroma or in small groups. The ascospores are allantoid, dark-brown, one-celled at first but become septate and very similar to those of *Cryptosphaeria.* In fact *Endoxyla fraxini* and *Cryptosphaeria eunomia* are synonymous.

Anthostoma: The genus *Anthostoma* (subgenus *Lopadostoma*) also shows relationships to several larger taxa. *Anthostoma phaeosperma* and *A.* (*Quaternaria*) *morthierii* have allantoid ascospores which suggest their origin from the Diatrypaceae. On the other hand, other species of *Anthostoma* have inequilateral dark-brown one-celled ascospores similar to those of *Rosellinia* and the Xylariaceae. Von Arx and Müller (**1954**) place both *Anthostoma* and *Lopadostoma* in the Xylariaceae. The genus seems to include species of of two different origins.

Valsaria: The genus *Valsaria* also includes a heterogeneous group of species and their ordinal position is open to question. *Valsaria moroides* and *V. allantospora* have ascospores which are cylindric to allantoid, one-celled at first with two to six yellowish areas and constricted in the middle, which later become uniseptate. There is merely a blackening of the surface bark and periderm where the ostiolar necks are collectively erumpent. These species show a similarity to *Endoxyla* and the Diatrypaceae. *Valsaria insitiva* and *V. exasperans* have dark brown ellipsoid ascospores which are constricted at a central septum. These species have a strongly developed entostroma of dark colored fungus tissue which is reddish in *V. rubricosa*. Conidial locules with ellipsoid conidia are found in these species. These characters suggest relationships with *Endothia* and the Diaporthaceae. Müller and von Arx (**1962**) place these species in the Diatrypaceae.

Eutypa: The genus *Eutypa* has effuse, eutypoid ascostromata, found mostly on decorticated wood. The perithecia are scattered singly and separately erumpent through the blackened surface of the wood or substrate surface, the blackening caused by the proliferation of dark colored hyphae. In species which have a scanty entostromatic development in the wood, allantoid conidia may be formed on short dark colored conidiophores arising from the surface mycelium. In other species there is a greater stromatic development and the entostroma may be raised above the surface and the effuse stromatic area may be outlined by a blackened zone within the wood. In such species, small cushions of ectostromata may be formed on the surface, within which allantoid to filiform conidia are formed within an enclosed cavity. In *E. spinosa* on weathered oak wood the emergent ostioles form a blackened surface crust.

Cryptosphaeria: This is similar to *Eutypa* in having effuse eutypoid ascostromata, but usually occurs on corticated wood and forms a blackened layer of hyphae beneath the periderm, corresponding to the surface blackening in *Eutypa*. The periderm is not ruptured except where the ostioles are erumpent through it. No ectostroma is formed and the conidial state, of the form genus *Cytosporina*, is formed in labyrinthiform locules within the entostroma which is outlined by a blackened zone within the bark. The conidia are filiform, curved and hyaline. Small allantoid conidia may also occur.

Cryptosphaeria eunomia (*millipunctata*) has cylindric to allantoid ascospores which are hyaline when young, but dark yellow-brown and one- to four-septate at maturity, similar to those of *Endoxyla*, indicating a relationship to that genus.

Eutypella: In *Eutypella* the stromata are valsoid or pustulate with the perithecia in distinct groups and the ostioles collectively erumpent through the periderm as a disc. The blackened marginal zone, however, may be continuous from one valsoid group to another. The ostioles are often elongate and sulcate at the apex. Small cushions of ectostroma are first formed just beneath the periderm and function to rupture this tissue. Labyrinthiform conidial locules arise either within the ectostroma or the entostroma, within the bark. The conidia are allantoid to filiform, hamate and extruded in an orange or yellow spore horn or mass.

Quaternaria: This is an ill-defined genus. The stromata are valsoid, but there are supposed to be only a few (one to four) perithecia in each group. Most species can be placed in other genera.

Diatrype: In this genus the stromata may be disc-shaped and limited in area with a sharply defined margin, or widely effuse (as in *D. stigma*). *Diatrype stigma* is a common species on the bark of various hardwoods. A thick effuse layer of hyaline ectostroma is formed on the surface of the bark beneath the periderm. At various points this layer increases in thickness to form stromatic cushions within which labyrinthiform conidial locules arise. The conidia formed in these cavities are crescent-shaped, one-celled, hyaline and are exuded in yellowish spore horns. This ectostroma increases in thickness over wide areas and throws off the periderm exposing a powdery brown surface of the entostroma, within the bark. This surface layer turns brown-black and numerous crowded perithecia originate within the entostroma which is outlined by a blackened zone. The entostroma is, therefore, erumpent above the periderm and the perithecial ostioles are separately erumpent through its surface. In other species, as *D. disciformis*, the erumpent entostroma is limited in area to a circular disc with a sharp margin. In *D. virescens* the entostroma is green in color. The genus differs from *Eutypella* in the separately erumpent ostioles and the sharply margined discs.

Diatrypella: In this genus the asci contain numerous small allantoid ascospores. The stromata are valsoid but the marginal blackened zones may enclose more than one group of perithecia. The ectostroma in this genus is poorly developed and does not throw off the periderm. As a result the periderm is more adherent to the small disc of collectively erumpent ostioles. The conidial cavities may occur in the ectostroma or extend into the entostroma, in the bark. The conidia are, again, allantoid to filiform, hamate and yellow to orange in mass. Several species of *Diatrypella* have a greenish entostroma.

Cryptovalsa: The genus *Cryptovalsa* is very similar to and grades off into the genus *Diatrypella*. The entostroma in *Cryptovalsa* is more effuse and includes a number of small valsoid groups of one to four perithecia, with ostioles which are collectively erumpent. The asci are polysporous and the ascospores are allantoid and yellowish.

Rhynchostoma: This genus was placed in the Diatrypaceae by Müller and von Arx (**1962**). The perithecia are erumpent on a slight hypostromatic development beneath the host (wood) surface. The ostioles are elongate.

The ascospores are two-celled, brown and ellipsoid, not allantoid, as in the other genera of the family.

Xylobotryum: This genus is intermediate between the Diatrypaceae and the Xylariaceae and illustrates a possible relation between these two families. The perithecia are borne on an erect clavate or lobed stroma as in *Xylaria*, but the asci do not have the spot in the ascus tip which is blue with KI. The ascospores are fusoid and brown as in the Xylariaceae, but they are two-celled and without the germ slits characteristic of that family.

CORONOPHORACEAE:

In this family the perithecia are turbinate or globose and are grouped or gregarious, immersed or in a surface subiculum or on a hypostroma. They have no ostioles but open widely at maturity. The asci are clavate with a more or less elongate stalk and are very numerous in a basal fascicle. An inner hyaline layer of the perithecial wall is thickened above and consists of gelatinous cells which swell at maturity as a "Quellkörper" which extrudes the asci in a compact ball. The family has been placed in the Plectascales on account of the lack of an ostiole but the asci and the perithecial centrum are entirely different.

I. Asci polysporous
 A. Ascospores 1-celled
  1. Ascospores allantoid
   a. Asci long-stalked, perithecia immersed, with Quellkörper ....
   .... CRYPTOSPHAERELLA
   b. Asci short- or long-stalked; without Quellkörper .... CORONOPHORA
 B. Ascospores allantoid with divided protoplast .... FRACCHIAEA
II. Asci 8-spored
 A. Ascospores 1-celled, hyaline
  1. Ascospores with terminal appendages .... BICILIOSPORA
  2. Ascospores without terminal appendages .... CORONOPHORELLA
  .... (SCORTECHINIA)
 B. Ascospores 1- to 2-celled, brown .... GAILLARDIELLA

Cryptosphaerella: This genus has small clavate asci, with a long thin stalk, containing numerous allantoid ascospores. The asci are borne in dense fascicles on a basal cushion of hyaline parenchyma. The upper portion of the inner wall of the perithecium consists of hyaline cells which swell to form a Quellkörper which ruptures the non-ostiolate outer wall.

Coronophora: This has urn-shaped, flat-topped perithecia. The asci are polysporous and have short stout stalks. There is no Quellkörper and the asci are released in a compact ball.

Coronophorella (*Scortechinnia*): This genus has turbinate perithecia borne in a superficial subiculum, which has spine-like setae in some species. The asci are small, eight-spored, short stalked and borne in a basal fascicle. A Quellkörper is formed in the upper inner wall. Star shaped setae occur on the subiculum.

Scortechinniella: This genus also has flat topped perithecia borne on a subiculum. The asci, however, are polysporous, long-stalked and borne on a basal cushion beneath an apical Quellkörper. The ascospores are elliptical, flattened on one side and have a filiform appendage at each end.

Fracchiaea: In *Fracchiaea* the perithecia are grouped on a subepidermal stroma, but are broadly erumpent. The inner Quellkörper layer ruptures the inner wall at the apex forming a pore, but the outer dark brown wall remains intact. The asci are small, long-stalked, polysporous. The ascospores are allantoid with a divided protoplast giving the appearance of a two-celled spore. The asci are on a basal cushion.

Biciliospora: This genus has erumpent flat-topped perithecia on a subiculum. The asci are small, long-stalked and contain eight ellipsoid ascospores which have a filiform appendage at each end.

Galliardiella: This genus has erumpent flat-topped perithecia with the side and basal walls much thickened. The asci are small, long-stalked and eight-spored and borne in a basal fascicle. The ascospores are one-celled and hyaline at first, becoming one-septate and brown at maturity.

CALOSPHAERIACEAE:

The perithecia of the Calosphaeriaceae have a short, stout ostiole. They are grouped in loose clusters or scattered just beneath the periderm, which is ruptured in an irregular manner. In some cases there may be a slight development of stromatic hyphae about the perithecia. The asci are small, numerous and in a dense cluster in the base of the perithecium. The ascospores are allantoid and paraphyses are present. The species are inconspicuous on woody stems.

I. Asci many-spored .... PLEUROSTOMA
II. Asci 8-spored
    A. Ostioles short
        1. Perithecia tomentose .... ENCHNOA
        2. Perithecia smooth
            a. Perithecia disc-like or pezizoid .... ROMELLIA

b. Perithecia globose
    (i). Perithecia valsoid .... EROSTELLA
    .... (TOGNINIA)
    (ii). Perithecia scattered .... JATTAEA
B. Ostioles elongate
    1. Perithecia valsoid .... CALOSPHAERIA
    2. Perithecia scattered .... WEGELINA

Pleurostoma: This genus has long stalked, many-spored asci and clustered perithecia. The ostioles are sometimes laterally placed.

Enchnoa: This genus has immersed, tomentose perithecia with flattened ostioles. The asci are rather large and long-stalked.

Romellia: This genus has smooth perithecia with disc-like ostioles and asci clustered on long stalks.

Erostella: This genus has perithecia borne in valsoid clusters with short wart-like ostioles.

Jattaea: This genus has scattered perithecia with wart-like ostioles.

Wegelina: This genus has scattered perithecia with elongate ostioles.

Calosphaeria: This genus has valsoid clusters of perithecia with elongate ostioles.

NITSCHKIACEAE:

Fitzpatrick (**1923**) erected the sub-family Nitschkieae but was unable to suggest its relationship in any described family. The species are all saprophytic on decaying wood. The perithecia are turbinate with a short basal stalk. They occur thickly scattered on a surface subiculum which has a metallic iridescence, or on a thin parenchymatic stroma. In age, they collapse apically in a pezizoid manner. The ascospores are various. The asci are clavate and long-stalked.

I. Perithecia setose, on a subiculum with spines .... ACANTHONITSCHKEA
II. Not setose as above
    A. Ascospores hyaline
        1. Ascospores allantoid, 1-celled .... NITSCHKIA
        2. Ascospores filiform, 2-celled .... CALYCULOSPHAERIA
    B. Ascospores colored
        1. Ascospores 1-celled .... TYMPANOPSIS
        2. Ascospores 3-septate .... THAXTERIA

Nitschkia and *Acanthonitschkia* have allantoid ascospores. In *Nitschkia* the perithecia are borne on a basal stroma, in clumps and are erumpent through the bark. In *Acanthonitschkia* the perithecia are seated on a subiculum and sharp pointed setae occur on both the subiculum and the perithecia.

Calyculosphaeria: This genus was placed in the Coronophoraceae by Müller and von Arx (**1962**). Fitzpatrick, however, described and figured the genus as having papilliform ostioles and pezizoid collapsing perithecia. The ascospores are fusoid, hyaline and one-septate with pseudosepta.

Thaxteria: In *Thaxteria*, the perithecia are coarsely tuberculate and shiny. The ascospores are three-septate and brown.

Tympanopsis: This genus has one-celled brown ascospores.

DIAPORTHALES

The Diaporthales is an order with rather vague limits. Höhnel (**1917**) erected the Diaportheen as a family and segregated the Valsaceae as a second family. The asci, according to Höhnel, are clavate, have a short evanescent base and are freed within the perithecial centrum as a mass without any definite arrangement (i.e., not as a definite hymenium). The asci are clavate and have a thickened apical wall penetrated by a small canal which is surrounded by a refractive ring, appearing as two refractive spots in section. The perithecia are immersed and the ostioles in many species have a meristematic growing point and may become elongate and project above the substratum. Müller and von Arx (**1962**) have included some genera (*Phomatospora, Sydowiella* etc.) with parallel asci and a definite hymenium and others (*Hercospora*) with a different arrangement of the apical apparatus in the ascus. Species in which the asci remain parallel have an ascus with a short stout base which is held in position by the presence of numerous paraphyses.

The perithecia may be scattered singly or grouped in various types of stromata. Intermediate types do occur, but the great majority of species show either the presence or absence of a stroma. The non-stromatic forms are, therefore, placed in a separate family, the Gnomoniaceae. Müller and von Arx (**1962**) place all genera in one family, the Diaporthaceae.

The Japanese species of the family Diaporthaceae (including the Gnomoniaceae) have been treated by T. Kobayashi (**1970**).

I. Ascospores allantoid, hyaline .... VALSACEAE
II. Ascospores not allantoid
  A. Perithecia scattered or grouped, within an entostroma, or beneath an ectostroma .... DIAPORTHACEAE
  B. Perithecia scattered singly, not within any stroma .... GNOMONIACEAE

GNOMONIACEAE:

In this family the perithecia are borne singly within the substrate without any true stromatic development. Some species have a tendency toward such a stromatic development by the formation of dark colored hyphae around the ostiole forming a disc known as a "clypeus." Most species occur on leaves or herbaceous stems, a few on woody stems. The ostioles are often elongate and erumpent.

I. Ascospores 1-celled
  A. Asci 16-spored; ascospores finally with 1-3 septa
    1. Perithecia singly erumpent through an ectostroma .... DITOPELLA
    2. Perithecia erumpent as an elongate ostiolar neck .... REHMIELLA
  B. Asci 4- to 8-spored
    1. Perithecia erumpent as a flat ostiole through an epidermal clypeus ....
       .... SPHAEROGNOMONIA
    2. Without such a clypeus
       a. Ostiole short; asci parallel in a hymenial layer .... PHOMATOSPORA
       b. Ostiole erumpent, elongate; asci lying free at all levels in the perithecium
          .... GNOMONIELLA
II. Ascospores 2-celled (not filiform)
  A. Ascospores apiosporous; septate below or above the middle
    1. Perithecium lying horizontally; ostiolar neck lateral
       a. Asci lying parallel in a hymenial layer .... PLAGIOSTIGME
       b. Asci lying free, at all levels...................... PLAGIOSTOMELLA
    2. Perithecia vertical; ostiolar neck apical
       a. Ascospores brown, septate below the middle; perithecia erumpent through a subepidermal clypeus; asci lying parallel in a hymenial layer
          .... ANISOMYCES
       b. Ascospores hyaline, septate below the middle; asci lying free at all levels; no clypeus .... APIOGNOMONIA
  B. Ascospores double spindle-form, strongly constricted at the middle
    1. Perithecia lying horizontal; ostiolar neck lateral
       a. Asci lying parallel in a hymenial layer .... DIATRACTIUM
       b. Asci lying free at all levels .... PLEUROCERAS
    2. Perithecia vertical; ostiolar neck apical......................... VIALAEA
  C. Ascospores ellipsoid, neither apiosporous nor double-spindle-form
    1. Perithecia lying horizontal; ostiolar neck lateral .... PLAGIOSTOMA
    2. Perithecia vertical; ostiolar neck apical
       a. Asci lying free at all levels; ostiolar neck elongate............ GNOMONIA
       b. Asci lying parallel in a hymenial layer; ostiolar neck stout, erumpent ....
          .... SYDOWIELLA
III. Ascospores filiform
  A. Perithecia horizontal; ostiolar neck lateral
    1. Perithecia in a pseudostroma in a swelling of the leaf, surrounded by a blackened zone on both leaf surfaces; ostiolar neck elongate; ascospores non-septate... LINOSPORA
    2. Perithecia not in a pseudostroma; ostiolar neck short; ascospores with indistinct septa at maturity .... PLAGIOSPHAERA
  B. Perithecia vertical; ostiolar neck apical
    1. Ascospores filiform, with septa .... OPHIOCERAS
    2. Ascospores with an ellipsoid upper and a long filiform lower cell ....
       .... OPHIOGNOMONIA
    3. Spores filiform with indistinct septa................. GAUEMANNOMYCES

Ditopella: The genus *Ditopella* is intermediate between the stromatic and non-stromatic types of the Diaporthales. The perithecia are immersed singly in the bark beneath a well developed ectostroma which ruptures the periderm. The single ostiole is erumpent through this ectostroma. The asci lie at all levels, have a refractive ring in the apex and contain sixteen one-celled ascospores. The ascospores are ellipsoid, one-celled or finally septate with two to three droplets. The fungus has the structure of *Cryptosporella* but the perithecia are separately erumpent.

Rehmiella: In this genus the asci are also sixteen-spored, but the ostiolar necks are elongate and penetrate the epidermis without any ectostroma. The ascospores are one-celled.

Sydowiella: In this genus the perithecia are scattered singly in the substrate, with elongate ostiolar necks, as in *Gnomonia*, but the asci are cylindric and attached in a definite hymenial layer and the ascospores are two-celled, ellipsoid rather than fusoid as in *Gnomonia.*

Phomatospora: This genus has one-celled ascospores, uniseriately arranged in elongate cylindric asci which remain attached in a definite hymenial layer, but with a faint refractive ring in the apex. The perithecia are found scattered singly on herbaceous stems and grass culms.

Sphaerognomonia: This genus has small perithecia with short ostiolar necks, scattered singly in leaves with a clypeus-like blackening about the ostioles. The asci lie at various levels and contain one-celled ascospores.

Plagiostigme and *Plagiostomella*: Both genera have perithecia which lie horizontally, parallel to the host (leaf) surface and have lateral ostiolar necks which bend upward at an angle of 90° to penetrate the leaf surface. Both genera have ascospores with the lower cell smaller than the upper one (apiosporous). *Plagiostigme* has asci which are permanently attached in a definite hymenial layer, whereas *Plagiostomella* has asci which lie free in the perithecial cavity (see *Plagiostoma*).

Plagiostoma: This genus is used by von Arx and Müller for species with horizontal perithecia and equally two-celled ascospores. Cash and Waterman (**1957**) described *Plagiostoma populi* as causing a leaf disease of *Populus*. The ascospores in the latter species have the septum toward the lower end of the spore as in *Plagiostomella*. Höhnel originally erected the genus *Plagiostomella* for a species of *Plagiostoma* (*P. petiolicola*) which had unequally two-celled ascospores. Klebahn found this species to have the septum at the upper end of the ascospore. Höhnel, therefore, changed the type species to *P. carpinicola* which has the septum in the lower end of the spore. Von Arx (**1951**) considered the original type species (*P. petiolicola*) to be a *Plagiostoma*, because of the upper cell of the ascospore being smaller. He limited *Plagiostomella* to species with the lower cell being smaller. Cash and Waterman would place all species with horizontal perithecia and two-celled ascospores, whatever the position of the septum, in the genus *Plagiostoma.*

Diatractium and *Pleuroceras*: These two genera also have the perithecia lying horizontally, parallel to the host surface, but the ascospores in these two genera are double spindle-form, two-celled, hyaline and strongly constricted at the septum. In *Pleuroceras* they tend to separate into the two cells. *Diatractium* has asci which are firmly attached at the base and remain parallel in a peripheral hymenium, whereas *Pleuroceras* has asci which are soon detached from their base and come to lie free in the perithecial centrum. *Marssonia rhabdospora* is given as the conidial state of *Pleuroceras populi.*

Vialaea: This genus also has double spindle-form ascospores, but the cells are not separable and the perithecia lie in a vertical position with an apical ostiole. The tip of the ascus takes a blue stain with potassium iodide, relating this genus to the Amphisphaeriaceae.

Anisomyces: This genus has brown, unequally two-celled (apiosporous) ascospores with the lower cell being small papillate and lighter in color. The stout ostiolar neck is erumpent through a clypeus-like stromatic development.

Gnomonia: The genus *Gnomonia* has immersed perithecia, usually with elongate ostiolar necks. The asci have the characteristic refractive ring in the apex of this family. The species are mostly parasitic upon leaves. The genus originally included species with ascospores of various types, but several genera have been segregated upon ascospore characters and the genus limited to those species with equally two-celled ascospores. The conidial states are mostly formed in acervuli, but have been placed in a number of form genera.

Gnomonia leptostyla causes a leaf spot of *Juglans* (walnut). Its conidial state is a *Marssonina*, which has lunate two-celled conidia.

Gnomonia erythrostoma is the cause of a cherry leaf-spot. Its pathology and development has been studied by Frank (**1886**), Brooks (**1910**) and Likhite (**1926**). Infection occurs on the leaves in the spring. The mycelium passes into the petioles and prevents the formation of the abscission layer. As a result the leaves hang on the tree over the winter and cause reinfection in the spring. Spermagonia appear in large numbers in August as circular spots on the lower surface of the leaf. They form small acervuli covered by the epidermis. Long filiform, curved, or hamate spermatia are formed in large numbers, but these do not germinate. Perithecial initials appear soon afterward as coiled hyphae which become compacted into a ball, near the lower epidermis. Hyphae from these coils emerge and grow out through the stomata. These have been called trichogynes and supposedly fuse with one of the numerous spermatia. Brooks, however, saw such hyphae arise from ordinary vegetative hyphae and could not find fusion with the trichogynes nor any transfer of nuclei to take place. The perithecia mature from November to March. Several cells within the original coil become enlarged and take a deeper stain. These are the ascogonia, but they appear to disintegrate before the asci arise. The ascogenous hyphae appear in the base of the perithecium from vegetative cells. Broad band-like hyphae grow up into the perithecial cavity but are dissolved away as the asci develop. The asci arise from croziers. Likhite (**1926**) in his study of this species found that the perithecia arise as a definite coiled archicarp, as reported for *G. intermedia*. He also believed that the ascogonial cell in the archicarp was connected with an adjacent "antheridial" cell by means of pores in the walls and that fertilization took place by the migration of nuclei from one cell into the other.

Morgan-Jones (**1953**) studied the development of *Gnomonia intermedia*, a leaf spot pathogen of *Betula*, in pure culture. He observed the formation of ellipsoid, hyaline, one-celled conidia cut off successively from the sterigmata of lateral cells of the mycelium in culture, but found not pycnidia occurring in nature. The perithecia arose as a definite coiled hypha or archicarp of deep staining multinucleate cells. The apex of this archicarp was continued into a trichogyne which remained within the perithecial primordium, or extended beyond it. The sheath cells about the archicarp arose from adjacent vegetative hyphae. Pores were seen in the walls between the cells of the archicarp. Evidence of the fusion of the trichogyne with an "antheridial filament" within the primordium, was seen. Numerous filaments grew out from such a primordium (as in *Neurospora*). Some of these might have been trichogynes, others merely tomentum.

Single spore cultures were self-sterile and did not produce perithecia. Morgan-Jones called these homothallic, but crosses between them showed the typical bipolar heterothallic pattern. As the primordia develop, the ascogonium disappeared and the ascogenous hyphae appear as fragmented cells in the base of the cavity formed by the disintegration of the inner wall cells. The nuclei were not paired in the ascogenous hyphae and the asci arose from croziers or directly from lateral bi-nucleate cells of the ascogenous hyphae.

Apiognomonia: Apiognomonia is similar to *Gnomonia* but includes species which have ascospores with the septum in the lower end of the spore. *Apiognomonia veneta* (*A. errabunda* according to Müller and von Arx) is the cause of a leaf spot, or anthracnose of the leaves of sycamore

(*Platanus*) and other trees. On living leaves the conidial state is formed in an open acervulus (*Gloeosporium*) with ellipsoid, one-celled, hyaline conidia. On stems the conidial fruit body is immersed and enclosed and is placed in the form genus *Discula*. On dead leaves the conidial hymenium is enclosed within a pycnidial wall and falls in the form genus *Sporonema* (*S. platani*).

The black spot of elm is caused by a fungus known as *Gnomonia ulmea*. Actually, the ascospores of this species are apiosporous and it probably should be placed in *Apiognomonia*. Müller and von Arx (1962) placed this species in *Lambro*, as *L. ulmea*, but the perithecia, as shown in their figure (309) are not contained in a stroma. The development of this species has been described by Miles (1921) and Pomerleau (1938). A similar leaf spot, *Systremma ulmi*, occurs in Europe and was formerly confused with *G. ulmea*. The disease is common on the American elm and appears in early spring as small yellow leaf spots, on which numerous, minute, black stromata appear and may coalesce. These consist of a thin layer of dark colored parenchyma beneath the cuticle. A hymenium of short conidiophores is formed on the surface of this stroma and ruptures the cuticle exposing the mass of bacilliform conidia. The perithecia originate in May beneath these stromata. According to Miles, a deep staining hypha grows downward from such a stroma and coils to form an archicarp. He stated that other hyphae grow down around it to form the wall cells of the perithecium. Pomerleau, on the contrary, stated that the coiled archicarp arises first and that the trichogyne grows up from it and may penetrate the overlying stroma. No fertilization was noted. The wall cells arise from hyphae of the coiled archicarp. As the perithecium develops thick-walled hyaline to colored cells form an outer wall. The inner centrum is filled with thin-walled hyaline parenchyma. The coiled archicarp breaks up into segments. Several of these segments increase in size and have larger nuclei, which continue to divide. During August and September, these ascogenous cells divide to form binucleate cells from which the ascogenous hyphae arise. The asci arise by the formation of croziers and grow up into the centrum parenchyma which they disintegrate and use as a nutrient tissue. Meanwhile, a row of cells in the side of the perithecium toward the lower epidermis, become meristematic and give rise to the ostiole. The ascospores mature in December, the ostiolar necks do not rupture the overlying epidermis until the following spring, when the ascospores are ejected.

Gnomoniella: This genus is also similar to *Gnomonia* but has one-celled ascospores. *Gnomoniella tubiformis* occurs on leaves of *Alnus*. Its conidial state is *Leptothyrium alneum*. This genus differs from *Phomatospora* in the clavate asci with biseriate spores. The asci lie free in the perithecial centrum.

Ophiognomonia: This genus has elongate ascospores which are two-celled with the upper cell being narrow ellipsoid and the lower cell elongate filiform. The perithecia lie vertical with an apical elongate ostiolar neck.

Linospora: The perithecia of this genus lie horizontally and the elongate necks are lateral and bend at right angles to penetrate the epidermis of the leaf, through a clypeate development in the epidermis. The ascospores are long filiform and hyaline.

Miller and Wolf (**1936**) described *L. gleditsiae* as causing a leaf spot of locust. The conidial state has been described as a *Melasmia* or a *Leptostroma*. Miller and Wolf stated that the conidia are small, ellipsoid and formed on a thin sub-cuticular stroma covered by the blackened cuticle. They would place it in the form genus *Gloeosporium*.

Plagiosphaeria: This genus also has horizontal perithecia with short lateral ostiolar necks. There is no clypeus formed. The ascospores are long filiform and hyaline.

Gaeumannomyces: The pathogen causing the worldwide "take-all" or "white head" disease of wheat and other cereal grains has been known in the past as *Ophiobolus graminis* (*O. cariceti*). Jones (**1926**) has given the life history and development of this species in detail. He found that the asci have a refractive ring in the apex and referred the species to the Gnomoniaceae. The ascospores are long filiform and non-septate at first, but show pseudosepta at maturity. Von Arx and Olivier (**1952**), on the basis of Jones' work, created a new genus, *Gaeumannomyces* (*G. graminis*) of the Gnomoniaceae, for this species. Kobayashi gives *Gaumannomyces* as a synonym of *Linocarpon*.

The fungus is a soil organism and can be carried on the straw of the stubble and cause reinfection. It parasitises the roots and basal portion of the stalks where a mat of mycelium may be formed. Infection results in clumps of stunted plants with rotted roots and bleached heads of grain.

Jones found spermagonia arising as a clump of hyphae in the epidermal cells of the leaf sheaths of oats, in August. These primordia formed an ellipsoid pycnidium within which masses of small bacilliform spermatia were cut off from a peripheral hymenium. The perithecia originated about the same times as small coils (the archicarps) of cells, within the leaf cells. These coils became invested by neighboring vegetative hyphae, which gave rise to the perithecial wall. This archicarp coil branches and some of the cells enlarge and take on a deep stain. The tips of such branches were looked upon as trichogynes and the enlarged cells as ascogonia. The entire archicarp system eventually disintegrates, however, and remains functionless. Shortly after the archicarp appears certain vegetative hyphae grow into the perithecial primordium. These hyphae are lighter in their staining reaction and have better defined nuceli than those of the archicarp. The tips of these hyphae become swollen and break up to form a hyaline parenchyma in the centrum of the primoridium. Some cells of this central parenchyma, around the basal perimeter of the cavity develop a rich protoplasm and take a deeper stain. In these chains of cells there occur fusions as a result of the dissolution of the cross walls. Jones believed that these fusions may be between cells containing distantly related nuclei. The remaining parenchyma cells of the central cavity now become separated and elongate to form strips of cells oriented toward the ostiolar opening. The deeply staining cells give rise to ascogenous hyphae which develop in a disintegrating layer of cells on the inner portion of the lower wall. The asci arise at different levels and mature at different times. The spores were given as six-celled at maturity. There is a refractive ring in the apex of the ascus. When the ascospores are discharged, the ascus wall breaks down and the fascicle of spores is held together by the apical ring in the ascus and are forced out through the ostiolar neck.

This method of formation of interthecial tissues, as reported by Jones, indicates that a central mass of parenchyma can give rise to strips of interthecial tissue within the ascohymenial type of development, similar to that found in the Dothideales.

Von Arx and Olivier apparently overlooked the genus *Ophioceras* Sacc. (Syll. Fung. **2**: 328), which also has filiform septate ascospores. *Ophiognomonia* (*Gnomonia*) *padicola* also has filiform, two-celled ascospores, but not with the upper swollen and lower filiform cells of *Ophiognomonia.*

DIAPORTHACEAE:

This family shows many variations in the type of stroma in which the perithecia are imbedded. Most of the genera occur on woody stems, but a few are found on leaves. They may be saprophytic or parasitic. The ascospores are variable in shape, septation and color. In most genera the asci break away from their attachment early and lie at all levels in the perithecium, but in a few they are held in a hymenial layer by the formation and support of numerous paraphyses.

I. Ascospores 1-celled
A. On woody stems
1. Asci with deciduous stalks, lying at all levels, ascospores remaining 1-celled and hyaline
a. Perithecia without an ectostroma; brown layer of stromatic hyphae within the substratum .... GIBELLIA
b. Ectostroma present; entostroma of fungus plus host tissue
(I). Ascospores ellipsoid to fusoid .... CRYPTOSPORELLA
(II) Ascospores cylindric to elongate..................... CRYPTOSPORA
B. On leaves or herbaceous stems
1. Perithecia single or several in a well developed stroma of dark cells, 100-150 μ in diameter; on leaves .... DIPLACELLA
2. Stroma light colored, outlined by a blackened zone
a. Stroma on leaf hypertrophy
(I). Perithecia vertical; ostiolar neck apical, elongate...... MAMIANIELLA
(II). Perithecia horizontal; ostiolar neck lateral, short....... HYPOSPILINA
b. Stroma on herbaceous stems
(I). Entostroma effuse, of fungus and host tissue, containing many separately erumpent perithecia.................... DIAPORTHOPSIS
(II). Entostroma of fungus tissue only
(A). Perithecia borne singly within a blackened zone..... HETEROPERA
(B). Perithecia several in a flattened light colored stroma... MAZZANTIA
II. Ascospores 2-celled
A. Ascospores apiosporous
1. Ascospores angular, brown; stroma dark colored.............. PSEUDOTHIS
2. Ascospores ellipsoid, hyaline
a. Parasitic on leaves
(I). Perithecia horizontal; ostiolar neck lateral, short....... HYPOSPILINA
(II). Perithecia horizontal; ostiolar neck terminal
(A). Entostroma within leaf hypertrophy, surrounded by a blackened zone... MAMIANIA
(B). Entostroma erumpent; blackened zone on the outer surface only .... LAMBRO
b. Entostroma in woody stems
(I). Perithecia in a dark colored entostroma of fungus tissue............ .... ANISOGRAMMA
(II). Perithecia in a light colored entostroma of fungus and host tissue .... APIOPORTHELLA
B. Ascospores septate in the middle
1. Entostroma light in color, with a blackened outlining zone
a. Erumpent; on leaves; blackened zone on surface only ... PHYLLOPORTHE
b. On woody stems
(I). Asci short stalked to sessile, remaining parallel in a hymenial layer
(A). Perithecia thickly scattered in the stem hypertrophy; asci with refractive ring in the apex .... SAVULESCUA
(B). Perithecia in valsoid groups; outlined by a blackened zone; asci with a differentiated plug in the apex............. HERCOSPORA
(II). Asci deciduous at the base, lying at all levels; stroma valsoid, pustulate or effuse; asci with refractive ring in the apex............ DIAPORTHE
2. Entostroma light colored, containing fungus and host tissue, not outlined by a blackened zone; on woody stems
a. Conidial state a *Melanconium*; conidia dark colored, at least in mass; asci held in a parallel layer
(I). Asci with a refractive ring in the apex................. MELANCONIS

(II). Asci with a thickened apical wall, penetrated by a canal .... MASSARIOVALSA

b. Conidial state a *Septomyxa* with hyaline conidia in mass; asci at various levels with an apical ring .... CRYPTODIAPORTHE

3. Entostroma dark or bright colored

a. Entostroma yellow-brown, red or bright orange .... ENDOTHIA

b. Entostroma strongly developed, light colored within; surface blackened .... DIAPORTHELLA

III. Ascospores more than 2-celled

A. Asci 8-spored

1. Ectostroma present; entostroma light colored, poorly developed; ascospores often appendaged .... PROSTHECIUM

2. Entostroma dark colored, strongly developed

a. Ascospores light or dark colored, elongate, 3- or more septate; conidial state a *Coryneum* with superficial many-celled, dark colored conidia .... PSEUDOVALSA

b. Ascospores similar, with light colored end cells; without a *Coryneum* conidial state.................................... MELOGRAMMA

B. Asci with a single large ascospore .... TITANIA

IV. Ascospores muriform, with both transverse and vertical septa . PLEOMELOGRAMMA

Gibellia: In this genus the perithecia are imbedded in the wood and surrounded by layers of dark colored parenchymatic cells which also form a clypeus upon the surface. The ostiolar necks are erumpent and the ascospores are one-celled and hyaline.

Diplacella: The perithecia in this genus are small, 100-150 μ in diameter and scattered in the leaf tissue of leaf spots. A loose stroma of dark colored cells occurs as a thickened tissue on each surface of the leaf. The ascospores are one-celled and hyaline.

Mamiania - Mamianiella: In these two genera the perithecia occur singly or in groups within a light colored stroma of fungus tissues which is outlined by a blackened zone. In *Mamiania* (*M. fimbriata*) the ascospores are apiosporous and in *Mamianiella* (*M. coryli*) they are one-celled and hyaline.

Hypospilina: This genus has perithecia which lie horizontally within the leaf tissue and are surrounded by an outlining blackened zone. The ascospores are one-celled, ellipsoid, or two-celled with a septum toward the lower end.

Mazzantia: This genus forms light colored sclerotial stromata on herbaceaous stems, consisting of fungus tissue with a superficial blackened crust. The stroma may contain one or several perithecia with a dark colored diffferentiated wall. The ascospores are one-celled, ellipsoid and hyaline. Höhnel (**1918**) described the form genus *Mazzantiella* for the conidial state. It consists of a stroma similar to that of the ascospore state with a pycnidial locule containing comma-shaped conidia on filiform conidiophores. This is similar to the *Phomopsis* state of the genus *Diaporthe*.

Heteropera: This genus has single perithecia with a light colored wall, imbedded in a stromatic tissue of dark colored parenchyma. Höhnel considered this perithecium as a hyperparasite in the ascostroma of a *Leptosphaeria*.

Diaporthopsis: This genus has scattered perithecia immersed in an effuse entostroma consisting of mixed fungus and host cells and surrounded by a blackened zone. It is similar to an effuse *Diaporthe* on herbaceous stems, but has one-celled hyaline ascospores. *Diaporthopsis appendiculata* has ascospores with a long filiform appendage at each end. The perithecia may be scattered singly or in valsoid groups. The conidial states have been given as *Phoma* or *Phomopsis*. [Grove gave the conidia of *D. angelicae* as ellipsoid, hyaline and borne on cylindric arcuate "sporophores (beta conidia?)".]

Cryptospora - Cryptosporella: These two genera are found on woody stems. The perithecia are in valsoid clusters beneath an ectostroma through which the ostiolar necks are collectively erumpent. The entostroma is poorly developed and does not have an outlining blackened zone. In *Cryptosporella* the ascospores are ellipsoid, one-celled and hyaline. In *Cryptospora* the ascospores are cylindric to long filiform. In *Cryptospora*, the conidia are fusoid to fusoid-filiform, often curved, hyaline, one-celled and formed in enclosed cavities within the ectostroma; in *Cryptosporella* they are shorter fusoid. Smaller beta conidia may be produced in cultures of *Cryptospora*.

Petrak (**1966**) and T. Kobayashi (**1970**) used the name *Ophiovalsa* for *Cryptospora*, because of the prior genus *Cryptospora* in the Cruciferae.

Neoflageoletia: Saccardo described a subgenus *Flageoletia* of *Cryptosporella*, with one-celled ascospores, which was later raised to generic rank by von Höhnel. Reid and Booth (**1966**) found the type (*Flageoletia leptasca*) to be a *Phomatospora*, but described a new genus, *Neoflageoletia*, for the species *Flageoletia bambusina* which has both one-celled and two-celled hyaline ascospores. This genus has perithecia immersed within a dark colored stroma, and probably belongs in the Diaporthaceae.

The genus *Bagcheea* Müller and Menon (**1954**) is similar to *Cryptosporella* but it occurs on living leaves and the perithecia are imbedded in a stroma as in *Lambro*. The ascospores are hyaline to yellow-brown and one-celled with a large central vacuole dividing the protoplasm into two parts.

Pseudothis: This genus has subepidermal, erumpent reddish-brown, pulvinate stromata, within which several perithecia are imbedded, with short stout ostiolar necks. The ascospores are oblong to wedge-shaped with a septum in the lower end and are brown. The asci remain attached in a peripheral hymenium. The genus approaches *Valsaria* in some of its characteristics. Its conidial state has been placed in the form genus *Lasmeniella*. The conidia are disc-shaped, brown and with a light colored spot in the center of the flat side. They are borne in small cavities in a stroma.

Lambro: This genus has apiosporous ascospores with the septum in the lower end. The perithecia have light colored walls and are imbedded in a light colored stroma with a darker surface layer. Müller and von Arx (**1962**) placed *Gnomonia* (*Apiognomonia*) *ulmea* in this genus, but that species has no stroma.

Phylloporthe: This genus forms light colored stromata in the mesophyll of leaves. The stromata become erumpent, have a dark colored outer layer and contain several perithecia with erumpent ostioles. The ascospores are equally two-celled and hyaline.

Savulescua: This genus forms a hyaline effuse stroma within a swollen hypertrophy of stems. Such a stroma contains a large number of perithecia with separately erumpent ostiolar necks, and with a blackened outer crust. The ascospores are ellipsoid and two-celled and uniseriate in asci which are attached in a hymenial layer.

Anisogramma: This genus has unequally two-celled ascospores and a dark colored, well developed entostroma. Höhnel (**1917**), based his genus *Apioporthe* on *A. anomala*, which has such a dark colored entostroma and vertically elongated perithecia. Its ascospores are two-celled, the lower cell being small and cap-like. Müller and von Arx (**1962**) placed *Apioporthe*, therefore, as a synonym of *Anisogramma* and placed the remaining species of *Apioporthe* (as placed by Wehmeyer, **1933**) in the genus *Apioporthella*.

Apioporthella: this genus (*Apioporthe* of Wehmeyer) includes those species with unequally two-celled (apiosporous) ascospores and a poorly developed, light colored entostroma without any blackened outlining zone. The conidia, where known, are ellipsoid, one-celled, hyaline and formed in cavities within the ectostroma.

Diaporthe: This is a large genus. Most of the species are on woody stems, a few on herbaceous stems. Many of them are weak parasites attacking the host when it is injured or weakened in some way. The conidial states may occur on the living stems, the ascus state usually occurring later on dead or dying parts. The conidial states are of the form genus *Phomopsis* or related genera such as *Plenodomus* or *Fusicoccum*. Irregular conidial locules are formed in the ectostroma. The conidia are of two types. The first formed are the "beta" conidia (microconidia or stylospores) which are filiform, curved or hamate one-celled and hyaline. They do not germinate to form a mycelium. The "alpha" conidia (macroconidia) are ellipsoid or fusoid, one-celled, hyaline and do germinate. The ascus state is formed in an entostroma which may be effuse, pustulate or valsoid, and partially or entirely outlined by a blackened zone. The ectostromata are formed on the bark surface as small conic pustules which rupture the periderm and contain the conidial locules if such are formed. The ostiolar necks are elongate and singly or collectively erumpent through these ectostromatic discs. Some species as *D. eres* occur on a wide range of host substrata with little difference in their morphology. Others, as *D. decedens* on *Corylus, D. tessella* on *Salix* or *D. acerina* on *Acer*, seem to be limited to these host genera in nature. The ascospores are two-celled, hyaline and may be appendaged in some species.

Diaporthe phaseolarum causes a pod blight of lima beans and varieties of this species are parasitic on soy beans (*D. soyae*), sweet potatoes (*D. batatatis*) and other hosts. *D. citri* and its conidial state, *Phomopsis citri*, are the cause of a melanose of *Citrus*, causing black spots on the leaves, rot of the fruits and a "blast" of the twigs. The genus *Diaporthe* and related genera (*Cryptodiaporthe, Apioporthe, Diaporthella*) have been monographed by Wehmeyer (**1933**).

Cryptodiaporthe: This genus differs from *Diaporthe* in the absence of any blackened zone about the entostroma. The perithecia are in valsoid clusters and collectively erumpent. The conidial states are variable. Most of them have been placed in the form genera *Septomyxa, Fusiococcum, Discula*, etc. The conidia are one-celled ellipsoid, hyaline and sometimes with a faint septum at maturity. They are formed in locules in the ectostroma. *Cryptodiaporthe aubertii* var. *comptoniae* has two-celled, inequilateral to curved conidia which have a long filiform appendage at each end. This conidial state has been placed in *Neobarclaya* (*Pestalozzia*) *flagillifera*. In *C. aculeans*, the conidia are ellipsoid to fusoid, blackish hyaline (black in mass) and are produced on a sporodochium of the form species *Sporocybe* (*Stilbum*) *rhois*. The ascospores are two-celled and hyaline.

Diaporthella: This genus has valsoid groups of perithecia in a strongly developed entostroma which is widely erumpent through the periderm as a *Diatrype*-like disc. The ostiolar necks are collectively erumpent through this disc and are long and sinuous in some species. The perithecia are arranged in a polystichous manner in some species, and the blackened zone may or may not be present. The ascospores are two-celled and hyaline.

Melanconis - Prosthecium - Pseudovalsa These three genera are closely related and certain species of *Melanconis* show relationships with the other two genera and also to *Diaporthe, Massariovalsa* and *Valsaria*. The asci in these genera are more firmly attached at the base and are held in a hymenial layer by the numerous paraphyses which are formed. The three genera have been monographed by Wehmeyer (**1941a**).

Melanconis: The species of this genus are found on the stems of dicotyledonous trees, probably as weak parasites. They are commonly found on freshly cut or fallen limbs, or brush. The perithecial stromata are of the valsoid type with the ostiolar necks erumpent collectively through an ectostromatic disc. The ascospores may be either hyaline or brown and two-celled, although they may become several septate at maturity in a few species. The conidial states are of the form genus *Melanconium*. In this form genus the conidia are formed in cavities upon the flanks of the ectostroma. The alpha conidia are ellipsoid, one-celled and may be hyaline, in which case they are black in mass, or dark brown. Beta conidia, which are cylindric or often curved are found in these same cavities.

Certain species show transitional characters to other genera. In *M. thelebola*, the conidia are found in either enclosed locules or open cavities, and may become two- to four-celled at maturity. One-celled cylindric hyaline beta conidia are also formed. The ascospores are two-celled and hyaline but may become several septate and pigmented in age. These characters relate this sepcies to *Prosthecium*.

Melanconis corni has four-celled hyaline conidia formed in the ectostroma as in *Prosthecium* but has two-celled brown ascospores.

Melanconis modonia has a *Coryneum* conidial state with many-celled brown cylindric conidia borne on the surface of the ectostroma, as found in *Pseudovalsa*, but the ascospores are two-celled and hyaline, although they may become brown in age.

Kobayashi (**1970**) placed these species (*M. thelebola* and *M. modonia*) in the genus *Pseudovalsella* Höhn., because of the presence of many-celled brown conidia, as in *Coryneum*, the conidial state of *Pseudovalsa*. This is a good example of a common problem in taxonomic relationships. Intermediate species may often show characters of two different genera which have been arbitrarily separated upon such characters. The position of such species depends upon the opinion of the author in question.

Melanconis nigrospora and *M. thelebola* have perithecia imbedded in colored entostroma, outlined by a blackened zone and two-celled brown ascospores. These characters approach those of the genus *Valsaria*. *M. juglandis* is the cause of the death of branches of *Juglans cinerea* (butternut)(Graves, **1923**).

Massariovalsa was placed in the genus *Melanconis* by Wehmeyer (**1941**). The conidial state of this species belongs in the form genus

Melanconiopsis which has one-celled brown conidia as in *Melanconium*, but which are borne enclosed within the ectostroma. Müller and von Arx (**1962**) placed *Massariovalsa* as a separate genus, because of the gelatinous envelope about the ascospores and the somewhat different apical apparatus in the tip of the asci. There is also very little entostroma about the perithecia and the ostiolar necks are fused together.

Prosthecium: This genus includes species with perithecia grouped in a valsoid manner in a slightly developed light colored entostroma without an outlining blackened zone. The ascospores are three- to four(six)-celled, ellipsoid to cylindric, hyaline or brown and with a short or long appendage at each end. All species occur on dicotyledonous trees. The conidia are mostly in the form genus *Stilbospora*. The conidia are borne in cavities in the ectostroma and are ellipsoid to cylindric, several-celled, hyaline to brown.

Pseudovalsa: In *Pseudovalsa* the perithecia are grouped in a dark colored entostroma with a blackened surface zone. The ascospores are elongate ellipsoid to cylindric, many-celled and hyaline or brown. The conidial states fall in the form genus *Coryneum*. The conidia are elongate ellipsoid to clavate, many-celled, dark brown and are borne free on the surface of the ectostroma. Small cylindric beta conidia have also been reported.

Melogramma: This is a doubtful genus, similar to *Pseudovalsa*, with many-celled elongate cylindric brown ascospores, with light colored end cells. It is supposed to have conidia borne in cavities in the stroma. Doguet (**1960a**), in his study of *M. spiniferum*, found that this is an anomalous species with characteristics of both the Loculoascomycetes and the Euascomycetes. The perithecia are imbedded in a stroma and have differentiated walls and elongate ostiolar necks with periphyses and contain true paraphyses as in the Euascomycetes. However, the asci are bitunicate and eject the ascospores by the extrusion of a thickened inner wall. At the same time these asci have the refractive ring in the apex which is characteristic of the Diaporthales. The details of the origin of the perithecium are not known.

Titania: This is a monotypic genus based on a single collection. The type species *T. berkeleyi* has asci which contain only one single large many-celled ascospore. In other respects it is similar to *Pseudovalsa*.

Pleomelogramma: This genus is similar to *Melogramma*, but the ascospores are muriform with both transverse and vertical septa.

Endothia: In this genus the stroma is well developed and brightly colored, yellow-orange or red. The perithecia are numerous, deeply imbedded in the stroma, often polystichous and with elongate ostiolar necks erumpent to the surface. The ascospores are ellipsoid to allantoid and one- or two-celled. The conidial state is of the form genus *Endothiella*, which consists of numerous labyrinthiform locules in similar bright colored stromata. The conidia are numerous, allantoid one-celled and hyaline. The species occur on the roots and stems of trees. The genus has been monographed by Shear and Stevens (**1917**).

Endothia parasitica is the pathogen causing the chestnut blight, which has practically wiped out the chestnut in North America, and is also epidemic in Europe. *Endothia parasitica* is a native of Asia where it does little damage to the endemic Asian chestnut. It was introduced into the United States on nursery stock in 1913 and soon spread like wildfire throughout the eastern forests. At the present time there are practically no vigorous chestnut trees left in North America. Killed trees may sprout from the base but these sprouts are infected in turn and destroyed. Attempts to develop resistant strains have not been successful.

The fungus enters through wounds and the mycelium forms orange colored fans in the region of the cambium, causing sunken cankers eventually girdling and killing the branch or trunk on which it occurs. During wet weather, large numbers of conidial stromata of *Endothiella* are erumpent through the periderm. The numerous allantoid conidia ooze out in yellow slimy spore horns and are distributed by rain or on the feet of birds or the bodies of insects. Perithecial stromata are produced less abundantly during dry weather. This disease is a striking example of the increased virulence of a fungus introduced into a new area in which the available host species have not attained sufficient resistance to resist infection.

There are several species (*E. gyrosa, E. singularis*), which are endemic on *Quercus* and other members of the Fagaceae but do little damage in North America. Crosses have been made between North American and Asiatic species of *Castanea*, but the resulting hybrids have not produced trees of any value for timber.

VALSACEAE:

Genera of the Diaporthales with allantoid ascospores are placed in this family. These ascospores differ from those of the Allantosphaeriales in being entirely colorless in mass. The asci have a refractive ring in the apex and lie free in the perithecium. All the genera have a valsoid type of stroma and are found on woody stemms. The species are weak parasites causing twig or stem blights on weakened stems of woody plants.

The Czechoslovakian species of the family Valsaceae have been monographed by Urban (**1957**).

I. Conidial state phaeostilbaceous .... SCOPTRIA
II. Conidial state acervulate or pycnidial
  A. Conidial state acervulate .... GLOMERELLA
  B. Conidial state pycnidial
    1. Stromata outlined by a blackened zone
      a. Asci 8-spored .... LEUCOSTOMA
      b. Asci more than 8-spored .... VALSELLA
    2. Stromata not outlined by a blackened zone .... VALSA

Valsa: This is a large genus with many species occurring as weak parasites on the weakened branches of many woody plants. The perithecia are in valsoid clusters beneath a conic ectostroma which ruptures the periderm as a small whitish or grayish disc. There is very little or no entostromatic development and no blackened marginal zone. The asci are eight-spored, short stalked and lie free within the perithecium. The ascospores are allantoid, one-celled and hyaline in mass. The imperfect state is of the form genus *Cytospora*, consisting of irregular to labrynthiform cavities within an ectostroma. The small allantoid, hyaline conidia are formed in large numbers on conidiophores in a hymenium lining these cavities. The conidial pustules appear as small conic erumpent discs in large numbers on the living but weakened stems. The perithecial stromata usually occur later on the dead or dying branches. Many species have been described on the basis of host, but are not distinct morphlogically. *Valsa ceratophora*, for instance, has been shown, by Defago (**1942**), to have many morphologic synonyms on many different hosts. Urban (**1957**) recognized some twelve host varieties of *Valsa ambiens*. *Valsa sordida* and its conidial state, *Cytospora chrysosperma*, have been reported as the cause of a canker of poplars. Togashi (**1930, 1931**) has studied *V. japonica*, the cause of a canker and dieback of peach trees, and *V. mali* (**1924**) causing a canker of apple trees. Hemmi (**1916**) described *V. paulowniae* as the cause of a dieback of *Paulownia*.

Leucostoma: The species which are similar to *Valsa* but have a stronger developed entostroma which is bounded by a ventral blackened zone were originally placed in a subgenus and later in the genus *Leucostoma*. The conidial states have been placed in the form genus *Leucocytospora*. Kern (**1957**) has shown that this genus also has numerous strains which differ in their parasitism, host distribution, spore size and cultural characteristics, but are morphologically similar. As a result species delimitation is largely arbitrary. *Leucostoma kunzei* occurs on the stems and bark of *Pinus*, *Picea* and other conifers. *Leucostoma persoonii* (*L. leucostoma*) is found on *Prunus* and other members of the Rosaceae. It causes a canker of peach (Togashi, **1930, 1930a, 1931**) and cherry trees. *Leucostoma nivea*, with the conidial state of *Leucocytospora nivea*, is said to cause a canker of poplars.

Valsella: This is a genus with a stromatic development the same as that in *Leucostoma* but with asci containing more than eight ascospores. It is considered by some authors to be merely a polysporous form of *Leucostoma.*

Scoptria: The genus *Scoptria* was described for species with a valsoid to eutypoid stroma and allantoid ascospores, but with a phaeostilbaceous (*Harpographium*) conidial state consisting of dark colored, upright stalks bearing conidia. The type *S. isariophora* can occur without such conidial stalks and has been described as *Valsa heterocantha* and other *Valsa* species. Berlese gave this species as the type of this genus *Peroneutypella* which he described as having elongate ostioles and similar conidial state. The genus *Peroneutypa* has a similar conidial state but eutypoid perithecial stromata. These last two genera have allantoid ascospores which are colored in mass and long stalked asci lying at various levels in the perithecium. Although placed as synonyms of *Scoptria* in the Valsaceae, they may be more closely related to the Allantosphaeriaceae.

Glomerella: The genus *Glomerella* has generally been considered as falling within the Diaporthaceae. Von Arx and Müller (**1954**) placed the genus in their family Polystigmataceae. In many respects the development of the perithecia and the conidial states are similar to those found in the Gnomoniaceae. However, the asci remain attached in a definite hymenium and have a small refractive disc in the apex instead of the refractive ring characteristic of the Diaporthaceae. The perithecia in *Glomerella* occur immersed in the substratum, either scattered singly or clustered to form a pseudostroma. The ascospores are ellipsoid to allantoid, one-celled and hyaline. The conidia are borne in acervuli of the form genera *Colletotrichum* or *Gloeosporium*, and give rise to diseases known as anthracnoses.

A large number of species of *Glomerella* were described on many different hosts, but they have all been considered merely as varieties of *G. cingulata*, the cause of the bitter rot of apples.

As early as 1914, Edgerton noticed that certain strains of *Glomerella cingulata* produced perithecia in clumps whereas others formed scattered sterile perithecia and still others only conidia. When certain strains were crossed a ridge of perithecia occurred where the colonies met. The genetics of this species has been studied by a number of investigators (Edgerton, **1914**; Lucas et al., **1944**; Edgerton, Chilton and Lucas, **1945**; Chilton, Lucas and Edgerton, **1945**; Chilton and Wheeler, **1949, 1949a**; Wheeler and Driver, **1953**; Markert, **1949**; Wheeler and McGahen, **1952**; and Wheeler, **1950**). Without going into details of these many studies, it may be said that Chilton and Wheeler (**1949**, **1949a**) reduced the number of strains to four, which when crossed gave rise to a typical Mendelian behaviour for two pairs of characters, AB and ab, which are linked. The makeup of these strains is as follows: (It should be noted that the connotation plus and minus is not the same as that found in heterothallic species, which these strains are not.)

PLUS: Perithecia formed in glomerulate masses. Will mate with all three of the other strains. Genotype AB

MINUS: Perithecia scattered in small groups. Will mate with plus and conidial A strains. Genotype Ab

CONIDIAL A: No perithecia formed; conidia in large glomerate masses. Will mate with plus or minus. Genotype aB

CONIDIAL B: No perithecia formed; conidia in small masses. Will mate with plus only. Genotype ab

Gene A controls the formation of perithecia; gene a does not allow the formation of perithecia.

Gene B causes the clumping of perithecia or conidia; gene b gives rise to scattered perithecia or conidia in small acervuli.

Both A and B must be present in a zygote to obtain a line of increased perithecial formation.

In self-fertilized perithecia, all eight ascospores are of one parental type or the other. In hybrid perithecia (AB × ab) there is a segregation of four:four ascospores (Ab and ab or Ab and aB) or of two:two:two:two (as result of crossing over).

Perithecial development and the cytology of the ascus have been studied by McGahen and Wheeler (**1951**), Huttig, (**1935**), Lucas (**1946**) and Wheeler, et al. (**1948**). McGahen and Wheeler crossed a conidial B with a plus strain. In the self-fertile (plus) strain the perithecial initials arose as curved lateral branches from adjacent cells of the mycelium. The formation of one primordium appears to stimulate the formation of others in the immediate vicinity, giving clumped perithecia. In crossed cultures initials tended to arise where a hyphae of one strain crossed with another hypha (primordial initial) of a compatible strain, which might explain the formation of a line of perithecia between a fertile and a sterile strain. The two branches of an initial coil about one another to form an inner and an outer coil. The outer coil gives rise to the wall of the perithecium. The inner coil takes a deeper stain and develops into an archicarp of three to five cells. The tip cell of this archicarp abuts upon the wall hyphae of the outer coil and may penetrate through it. Occasional hyphae, seen growing out from the surface coil, have been interpreted as trichogynes, but these never function as such.

As this stage, in crossed cultures, hyphae of the opposite (conidial B) strain grow toward the perithecial initials and fuse with the tip cell of the archicarp of the initial of the plus strain, or may even grow through the wall hyphae of the initial to reach this tip cell within the outer coil. In single self-fertile cultures a similar fertilizing hypha arises from a neighboring vegetative hypha and fuses in a like manner. Shortly after such a fusion, the tip cell of the archicarp becomes binucleate and the primordium proceeds to maturity. Genetic evidence supports the view that there is a transfer of nuclei from one strain to the other. There is one original pair of nuclei in the tip cell which gives rise by mitosis to all other nuclei in the ascogenous hyphae, which arise by the lobing and branching of this tip cell. This type of fertilization recalls a similar situation reported by Jones (**1926**) for *Ophiobolus* (*Gaeumannomyces*) *graminis*, and by Morgan-Jones for *Gnomonia intermedia*. Asci are formed from croziers in typical fashion.

This is not a case of heterothallism, as in *Neurospora.* The A factor is necessary for the formation of perithecial initials (archicarps) and can also give rise to fertilizing hyphae. The B factor causes clumping of the perithecia and conidia. The conidial strains (without A) can give rise to fertilizing hyphae but not to archicarp formation.

ERYSIPHALES

The order Erysiphales conists of two families, the Erysiphaceae and the Meliolaceae, which are ecologically similar but not necessarily genetically related. They are both made up of superficial obligate parasites of higher plants and have globose ascocarps with an ascohymenial development. Together with the family Microthyriaceae, they were formerly placed in the order Perisporiales which was characterised by perithecia without any ostiolar opening. This ordinal name was abandoned because of the nomenclatorial confusion of the genus *Perisporium.* The Microthyriaceae have flattened or dimidiate ascocarps with various types of dehiscence, and for the most part have a stromatic type of development, placing them with the Bitunicatae. The Erysiphaceae have superficial globose perithecia without any ostiolar opening and with various types of appendages. They have a hyaline vegetative mycelium. The ascocarps resemble those found in the Plectascales, but differ in the definite basal hymenial cluster of asci. The Meliolaceae have superficial globose perithecia which often have a small but inconspicuous ostiole and a dark colored vegetative mycelium. Their relationships seem to be with the Microthyriaceae, but the ascocarps have an ascohymenial development, and unitunicate asci. These relationships need further study in view of the modern theories as to ascocarp development.

ERYSIPHACEAE:

The Erysiphaceae is a well defined family of obligate parasites causing important plant diseases of many higher plants. Its members are common in both temperate and tropical regions, but the perithecial state is seldom formed in warmer climates. The vegetative mycelium is colorless, almost entirely superficial and appears as a white moldy growth on the surface. Nutriment is obtained from the host by means of short lateral hyphal branches, or haustoria, which penetrate the surface host cells. The host tissues are not killed but photosynthetic activity is interfered with and early loss of foliage or destruction of fruit may occur.

During the growing season, a conidial state is formed which is similar in all species. Upright conidiophores arise from the superficial mycelium and cut off from their apex, by constriction, a succession of one-celled, ellipsoid, hyaline conidia. This results in a white powdery growth which gives the name powdery mildews to this group of fungi. This conidial state is similar in all species and was formerly known as *Oidium erysiphoides*. Hardison and others have recognized minor differences, however. In the genus *Phyllactinia* each conidiophore produces only a single conidium and such types have been placed in the form genus *Ovulariopsis*. *Leveillula* differs from the other genera in the formation of knots of hyphae in the stomatal chambers, which give rise to a cluster of conidiophores, each with a single elongate conidium. Such species are placed in the form genus *Oidiopsis*. This type of conidial state is similar to that found in some Parodiellaceae, except that there the vegetative mycelium is colored. The ecology of both families is similar.

The failure to form the ascigerous state in the tropics and warmer areas makes it difficult to determine species in such regions. The formation of perithecia seems to be a response to a change in the metabolism of the host. The perithecia in some species are formed as the host goes into a period of rest, and mature over winter on the dead plant parts, saprophytically. In the tropics where there is no common leaf fall or rest period, the fungus can be carried from one season to the next by the conidial state and the ascigerous state has been dropped out.

The perithecia originate in the late summer or fall, at the end of the growing season as minute orange colored granules on the leaf surface. Because of this pigmentation, Arnaud placed the family in the Hypocreales. The perithecia mature over winter on the dead plant parts and the ascospores mature and are distributed the following spring. The asci are saccate to broadly clavate and contain usually less than eight ascospores which are ellipsoid, one-celled and hyaline in the typical members of the family. The ascocarp has no ostiolar opening and in this respect resembles that of the Plectascales. The asci, however, are few, large and borne in a definite hyphal clump in the base of the "perithecium."

The perithecia have various types of appendages arising from the cells of the wall, which are useful as generic diagnostic characters. These appendages are probably instrumental in the distribution of the ascocarps and their contained ascospores. In such genera as *Erysiphe*, with undifferentiated appendages, the wall cells are similar throughout and there is equal shrinkage of the walls and the ascospores are distributed merely by the decay of the walls. In genera with differentiated appendages, the perithecial wall is dorsi-ventral, the basal cells having thin walls and the upper cells thicker walls. As a result the lower half dries in a concave manner, and repeated moistening and drying tears the perithecium from its attachment, when it may be blown about by the wind.

The family has been monographed by Salmon (**1900**) and Blumer (**1933**). A large number of species have been described on the basis of host occurrence, but many of these are morphologically identical and have been placed in synonymy, although physiological host strains no doubt occur.

The genera are separated upon the basis of the appendages and other characters, as follows:

- I. Appendages hypha-like, undifferentiated
  - A. Spores one-celled
    - 1. Perithecia with a single ascus .... SPHAEROTHECA
    - 2. Perithecia with several asci
      - a. Mycelium entirely superficial, conidial state *Oidium*.......... ERYSIPHE
      - b. Mycelium penetrating the leaf; conidial state *Oidiopsis*...... LEVEILLULA
  - B. Ascospores 2-celled.......... CHILEMYCES
  - C. Ascospores 4-celled.......... LEUCOCONIS
- II. Appendages dichotomously branched at the apex
  - A. Ascospores 1-celled
    - 1 Perithecia with a single ascus.......... PODOSPHAERA
    - 2. Perithecia with several asci .... MICROSPHAERA
  - B. Spores 2-celled .... SCHISTODES
- III. Appendages hamate, curved at the apex .... UNCINULA
- VI. Appendages needle-like with a bulbous base .... PHYLLACTINIA

The cytology of the development of the perithecium in the Erysiphaceae was among the first to be studied in the Ascomycetes, chiefly by Harper (**1895**), Dangeard (**1907**), Claussen (**1905-1912**) and others.

Sphaerotheca: The perithecia of this genus contain only one ascus and have undifferentiated appendages of ordinary hyphae. The development of the perithecium was first worked out by Harper (**1895**) for *S. castagnei* (*S. humuli*). The perithecia arise from two adjacent club-shaped uninucleate hyphal tips, called by Harper the oogonium and antheridium. The apical wall of the antheridium fuses with that of the oogonium (ascogonium) and forms an open passage with it, through which the antheridial nucleus migrates into the oogonium and fuses with the female nucleus (according to Harper). This diploid fusion nucleus then divides several times to form a row of cells, all of which are uninucleate, except the penultimate one, which is binucleate. This resembles the crozier formation in other Ascomycetes. This binucleate subterminal cell of the crozier arrangement then develops directly into the single ascus of the perithecium. Sterile hyphae grow out from the sexual branches and envelop them to form the wall of the perithecium. The appendages arise as outgrowths of the outer cells of the perithecial wall. The inner wall cells act as a nurse tissue for the developing asci.

This development is typical of the "sphaeriaceous" or ascohymenial type of development. Dangeard (**1907**), in his studies of the same species, reported a similar behavior, but he failed to see any open connection between the oogonium and the antheridium, and no fusion of nuclei within the oogonium (ascogonium). Winge (**1911**) in a study of the variety *fuliginea* of this same species also stated that there was no fertilization (fusion) within the oogonium.

It should be noted that the two nuclei of the binucleate cell which fuse within the ascus, according to Harper, are diploid as a result of the fusion within the oogonium, and therefore, (theoretically) the fusion nucleus in the ascus would be tetraploid. This led to the theory of brachymeiosis, presented by Gwynne-Vaughan (**1922**) which hypothesized a double reduction division within the ascus, with a second "brachymeiosis" to reduce the chromosome number to the haploid condition in the ascospores. This supposition colored all of the early investigations upon the development of the ascocarp within the Ascomycetes. The theory has been rejected, however, inasmuch as modern genetic studies have failed to find any evidence of a second reduction in the ascus.

Sphaerotheca humuli occurs on many different hosts and causes a disease of strawberries and hops. *Sphaerotheca mors-uvae* attacks the leaves and young berries of currants and gooseberries. It is a native of America but was introduced into Europe where it became very destructive. It causes a persistent layer of white mycelium on the leaves and fruits in May and June, which prevents the maturation of the fruit. Such a cottony mycelium is characteristic for the genus. *Sphaerotheca pannosa* is the rose mildew which causes woolly patches in the axis of the leaves and on the flowers and fruits, causing much damage in greenhouses. A variety *persicae* is found on peach trees.

Erysiphe: Erysiphe has undifferentiated appendages as in *Sphaerotheca*, but there is more than one ascus within the perithecium. The cytological development of *Erysiphe polygoni* (*E. communis*) was studied by Harper (**1905**). This is an ubiquitous species occurring on a large number of hosts. Harper reported a swollen ascogonial and a two-celled antheridial branch in this species, which come into communication with each other. The antheridial nucleus, he states, migrates into the ascogonium and fuses with the female nucleus. This fusion nucleus then divides, forming a row of five to eight nuclei which are cut off into a row of cells, the penultimate one of which is binucleate. Branches grow out from this penultimate binucleate cell to form a system of ascogenous hyphae. Some six to eight intercalary cells with paired (diploid) nuclei within this system enlarge and form asci within which a (second) nuclear fusion occurs, giving a tetraploid nucleus from which the ascospores arise. Dangeard (**1907**) studying this same species and *E. cichoracearum*, found no fusion in the ascogonium and stated that the asci arise from other cells than the penultimate one of the ascogonial branch.

Yarwood (**1935**) suggested that *E. cichoracearum* was heterothallic, from the fact that the perithecia arose along a line of contact between two colonies on sunflower leaves. Allen (**1936**), in her studies of perithecial development of *E. polygoni*, confirmed this idea. She found that the hyphal branches of opposite strains came in contact and formed lobes or branches which fused by means of a connecting pore, through which a nucleus migrated from one branch into the other. She could not distinguish between a male and a female branch, however. The cell into which this migration took place fused with a number of surrounding cells to form a contorted mass of communicating cells. She believed that all this central tissue was potentially dikaryon in nature. The original fusion cell died. There was a radial orientation of the other central cells, which became the young asci. Allen emphasized the difficulty in interpreting the irregular structures in the developing perithecium which had been previously interpreted upon the basis of the preconceived ideas in the literature. If the fusion of the cells or organs (called antheridia and ascogonia) be looked upon merely as an apparatus for the association and pairing of nuclei of differing genotype, much of this confusion disappears.

Erysiphe graminis is the cause of a disease of wild grasses and cereal grains which causes an enormous loss in these crops. Within this species there are many physiologic forms which differ in their ability to infect different groups of genera of grasses, but which cannot be distinguished morphologically. The groups of grasses which any strain will infect overlap, forming a confusion pattern. These differences, however, enable the geneticist to breed resistance to the various strains into desired host grasses, much as has been done in the rusts.

Podosphaera: This is a genus in which the perithecium contains only a single ascus and has appendages with an apical cluster of short dichotomous branches, each with a curved or coiled tip. *Podosphaera oxycanthae* is an ubiquitous species causing mildew diseases of many wild hosts, as *Amelanchier, Spiraea, Vaccinium,* etc.

Microsphaera: This genus has appendages the same as in *Podosphaera* but has more than one ascus in each perithecium. *Microsphaera alni* is the common species, occurring on many different hosts. It is particularly abundant on the leaves of lilac in the fall.

Uncinula: In *Uncinula* the perithecia contain several asci and the appendages have a single or once branched curved, uncinate tip. *Uncinula necator* causes a destructive disease of grape vines. The infected leaves shrivel and die. If the flowers are attacked, no fruits develop; if the fruits are infected they crack and die It appears to have originated in Japan. It spread to Europe about 1845, where it became a severe pest and spread to America about 1866.

Phyllactinia: In the perithecia of this genus, the appendages have a highly specialized structure. They occur as a circle of outgrowths just above the equator of the ascocarp. They consist of a long tapering, pointed apical portion and a bulbous basal portion. The upper side of the bulbous portion has a thicker wall than the lower side. Upon drying the lower wall dries more rapidly and wrinkles and contracts, thus pulling the setose portion of the appendage down with some force, raising the perithecium up on a circle of stilt-like legs. Alternate wetting and drying tends to tear the perithecium away from the mycelium. On the upper portion of the perithecium there are a number of tufts of short penicillately branched hyphae, whose walls form a gelatinous slime causing the perithecium to stick to anything it touches, thereby aiding in its distribution.

Phyllactinia corylea is a widely distributed species on such tree hosts as oak, beech, hazelnut, ash, etc. Its perithecial development was studied by Harper (**1905**) and is similar to that of *Erysiphe*. The ascogonial branch curves over the antheridial branch and receives a nucleus which fuses with a female nucleus. A row of three to five cells arises from the division of this fusion nucleus. The ascogenous hyphae arise from the penultimate multinucleate cell of this branch. The asci arise from binucleate cells of these ascogenous hyphae. Only two ascospores commonly mature in the ascus.

A more detailed account was later given by Colson (**1938**). According to this account, the two sexual branches are much alike They are at first uninucleate and coil about one another. The female branch soon becomes shorter, stouter and more curved. Each branch then cuts off a uninucleate tip cell. These apical cells divide into two cells in each branch. The apical cell of the male branch is small, that of the female branch enlarges. At about this time the sheath hyphae begin to branch off from the basal cells of the sexual hyphae and envelop them. There is no passage of male nuclei from the antheridium into the ascogonium, nor any nuclear fusion at this time, but the apical cell of the female branch becomes binucleate. The antheridial nucleus degenerates. The two ascogonial nuclei divide to form a row of four nuclei. Cross walls then form to give rise to an apical and a basal uninucleate and a median binucleate cell. The basal cell gives rise to the inner wall hyphae of the perithecium. The central binucleate cell elongates to form a curved multinucleate cell which gives rise to branching multinucleate ascogenous hyphae. The asci arise from binucleate cells cut off in these hyphae. These two nuclei fuse to form the diploid nucleus of the ascus. Upon meiosis of this fusion nucleus one large nucleus in the base and three small nuclei in the apex of the ascus are formed. All four of these nuclei divide. Astral rays are formed about the two large nuclei in the base of the ascus and two ascospores are cut out. The six supernumerary nuclei degenerate.

MELIOLACEAE:

The Meliolaceae consist of a group of superficial leaf parasites, widely distributed throughout the tropics and warmer climates. The genera included here were placed in the Perisporiales by Theissen and Sydow (**1917**), along with a number of other genera which have a similar habit of superficial parasites with globose perithecia. These other genera have been removed to such families as the Dimeriaceae and Perisporopsidaceae, in the bitunicate Ascoloculares, by Müller and von Arx. The Meliolaceae have globose superficial perithecia with inconspicuous ostioles, except for *Actinodothis* and *Amazonia*, which have flattened ascocarps and appear intermediate between this family and the Microthyriales (Höhnel, **1919**). The mycelium is entirely superficial, forming a dark colored crust leading to the common name of "sooty molds." Most species have short one- or two-celled lateral branches along the hyphae, known as "hyphopodia." These may be arranged oppositely or alternately on the hyphal strands. The apical cell of the hyphopodium may be swollen or "capitate," or tapered and "mucronate." There may be setae of various types on either the mycelium or the perithecia. The perithecia arise by the formation of a shield from the apical cells of the hyphopodium, beneath which the perithecium is formed.

A wide variety of conidial states has been reported for these genera, but there are numerous parasites or commensal fungi which grow with or upon the mycelial mats of these species and their true relationships are difficult to determine. Stevens (**1918**) has published upon these conidial associations in full.

These fungi cause leaf spots and normally do little damage to the host. The asci are cylindric to clavate, thin-walled and often evanescent. The ascospores are predominantly brown and two- to four-septate. Stevens (**1927-1928**) monographed the family. He segregated several genera on the basis of the presence or absence of various appendage types upon the perithecia. His key to the genera follows:

I. Perithecia dimidiate, flattened
  A. No free mycelium present .... ACTINODOTHIS
  B. Mycelium with hyphopodia present.............................. AMAZONIA
II. Perithecia globose
  A. Mycelium without capitate hyphopodia............................ MELIOLINA
  B. Mycelium with capitate hyphopodia present
    1. Mycelium without setae
      a. Perithecia with vermiform setae.................................. IRENE
      b. Perithecia with true setae (not vermiform).................... IRENOPSIS
      c. Perithecia with neither true setae nor vermiform appendages ..... IRENINA
    2. Mycelium with setae.. MELIOLA

Inasmuch as the abundance and character of the appendages upon the perithecium are often quite variable, even within the individual species, Doidge and Sydow (**1927**) would eliminate some of these genera. They gave the following simplified key to the genera:

I. Perithecia dimidiate .... AMAZONIA
II. Perithecia globose
  A. Mycelium without capitate hyphopodia............................ MELIOLINA
  B. Mycelium with capitate hyphopodia
    1. Mycelium without setae .... IRENE
    2. Mycelium with setae... MELIOLA

Actinodothis: This genus has a circular radiate ascocarp with one or more separate locules. It has been placed in the Microthyriaceae and the Dothideales by other authors. The asci are two-spored and the ascospores are three-septate and brown.

Amazonia: This also has a circular, radiate ascocarp, and has previously been placed in the Microthriaceae. It, however, has hyphopodia upon the mycelium and what appears to be a true perithecial wall beneath the radiate shield. The asci have a single thin wall, and contain the ascospores which are five-celled and brown. The type species was described as a *Meliola* and the fungus is transitional between the Microthyriaceae and the Meliolaceae. This also raises the question as to the position of these two taxa in relation to their ascohymenial or ascoloculares development.

Meliola: This is a large genus with globose perithecia, and includes all species which have capitate hyphopodia and setae upon the mycelium. They are found on a wide variety of host plants. The genus has been monographed by Gaillard (**1892**) and Beeli (**1920**).

Ward (**1883, 1883a**), in his study of the development of *Meliola*, believed that the two cells of a capitate hyphopodium acted as antheridium and ascogonium in the origin of the perithecium. Graff (**1932**) studied the development of *Meliola circinans*. He found that the ascocarp originated from a capitate hyphopodium by the diagonal division of the apical capitate cell to form a flat circular plate of cells which acted as an outer shield. He described the origin of the sex hyphae as coming from the basal cell of the hyphopodium as an enlarged ellipsoid ascogonium and an elongate, somewhat spiral antheridium, but saw no actual fusion. This ascogonium developed into a short septate hypha of uninucleate cells. Fine sterile hyphae arose from the basal cell of the antheridium and grew around the ascogonium to form the perithecial wall. Meanwhile the upper flattened shield became two layers thick and grew downward at the outer edges and around under the developing perithecium, forming an enclosed globose shield. Ascogenous hyphae arose from the cells of the ascogonium. The inner walls of the perithecium formed an inner parenchyma and from the cells of the upper portions of this wall there arose outgrowths which became the periphyses and penetrated through the outer wall to form an ostiole. The asci arose from croziers and were thin-walled. After meiosis and a mitosis of the fusion nucleus the resulting eight nuclei were cut out in pairs within four spore initials. Normally, two of these spores matured into five-septate brown ascospores and the other two disintegrated. This type of perithecial development differs from that found in the Microthyriaceae only in the development of the shield entirely around the base of the perithecium giving a globose structure, rather than a flat fruit body with an exposed under surface.

Irenopsis: Stevens put species with true setae upon the perithecium and no setae on the mycelium in this genus. Doidge and Sydow included such species in a section of *Meliola*.

Irene: Stevens limited this genus to species with "larviform" appendages. These are conic to elongate outgrowths of the perithecial wall cells with curved or uncinate tips. Doidge and Sydow claim that all degrees of development of such appendages occur, from mere mammillate cells to elongate appendages and included all such species, without true setae on the perithecium in the genus *Irene*. This includes Steven's *Irenina*.

Irenina: This is Steven's genus for species with no larviform appendages nor setae on the perithecia, and no setae on the mycelium, but with capitate hyphopodia.

Meliolina: This genus includes those species in which the mycelium has no hyphopodia.

SPHAERIALES

The order Sphaeriales, as here designated, is an artificial one to accomodate a number of families which do not fit into any of the previous orders of the Euascomycetes. They do not necessarily represent a phylogenetic group. The old order Sphaeriales included those genera of the Pyrenomycetes which had dark colored perithecia possessing ostioles. The studies of Höhnel, Nannfeldt and Luttrell transferred many of these genera to the Loculoascomycetes or Bitunicatae. A number of other orders were segregated on sets of characters supposedly indicating genetic relationships. There remained a residue of genera which did not show a relationship to any of these orders.

These genera have been more or less arbitrarily arranged on the basis of morphologic or ecologic similarity into the following families:

I. Asci with a thickened disc in the apex which stains blue with KI .... AMPHISPHAERIACEAE
II. Asci without any such structure
  A. Marine fungi from salt water; ascospores mostly with appendages............ .... HALOSPHAERIACEAE
  B. Fungi not marine in habitat
    1. Saprophytes on dung; ascospores large, 1-celled............. SELINIACEAE
    2. Not on dung
      a. Mostly parasitic leaf fungi, with a clypeus or stroma .... POLYSTIGMATACEAE
      b. Not with any of the above characters; mostly superficial saprophytes..... .... SPHAERIACEAE

POLYSTIGMATACEAE:

The family Polystigmataceae is difficult to circumscribe. Höhnel (**1917**, **1918**) and Petrak (**1928**) in a series of papers pointed out that the genera *Phyllachora, Polystigma, Physalospora, Physalosporina, Bagnisiopsis, Diachora*, etc. are all closely related. Nannfeldt (**1932**) erected the family Polystigmataceae for such genera. Von Arx and Müller (**1954**) have enlarged the family to include genera which have previously been placed in various orders, such as the Dothideales, Hypocreales, Diaporthales, etc. The typical members of this family are parasites on the leaves and stems of various plants and have perithecia imbedded beneath a blackened clypeus or within a stroma. Such a stroma may be bright colored, as in the genera *Polystigma, Phyllocrea*, etc., which have been placed in the Hypocreales. *Phyllachora* was originally placed in the Dothideales and many species of *Physalospora* belong in that order. *Glomerella* is related to the Diaporthales and *Ophiodothis* is similar to the Clavicipitaceae.

The asci in this family are thin-walled with a thickened apical wall in which there is a refractive disc. In some genera this apical plug may stain blue with KI. Müller and von Arx (**1962**) have transferred these genera to their Amphisphaeriaceae. This family is a good example of the re-arrangement of genera, formerly in widely separated orders, upon a different group of diagnostic characters.

The following key to the genera is a modification of that of Müller and von Arx.

I. Ascospores 1-celled
- A. Perithecia borne singly, without a clypeus or stroma
  - 1. Ascospores large (over 20 μ), thick-walled, granular, mostly colored .... PHYSALOSPORA
  - 2. Ascospores smaller, thin-walled and pale, not granular .... PLECTOSPHAERA
- B. Perithecia within a stroma or beneath a clypeus
  - 1. Ascospores brown with a hyaline equatorial band
    - a. Stroma sub-epidermal .... PHAEOCHORELLA
    - b. Stroma superficial, becoming free.................. PSEUDOTTHIELLA
  - 2. Ascospores without a hyaline band
    - a. Ascospores with a tapered tip
      - (I). Stroma light colored, causing a hypertrophy....... UROPOLYSTIGMA
      - (II). With a dark hypophyllous clypeus .... SCHIZOCHORA
    - b. Ascospores without a tapered tip
      - (I). Stroma entirely immersed beneath the epidermis
        - (A). Stroma yellow to reddish, fleshy, without clypeus .... POLYSTIGMA
        - (B). With a blackened clypeus or stroma
          - (1). Stroma effuse, blue with KI, clypeus both above and below; asci from side walls of the perithecium .... DIACHORA
          - (2). Stroma not blue in KI; asci from base of the perithecium
            - (a). Perithecium immersed, ostiole not protruding to surface .. .. ISOTHEA
            - (b). Perithecia immersed, but with ostioles protruding to surface .... PHYLLACHORA
      - (II). Stroma erumpent through the epidermis
        - (A). Stroma effuse, barely erumpent
          - (1). Stroma flat; ascospores hyaline .... PHYLLEUTYPA
          - (2). Stroma gall-like; ascospores pale yellow........... LOHWAGIA
        - (B). Stroma strongly erumpent, from a hypostroma
          - (1). Perithecia borne singly in an erumpent stroma
            - (a). Perithecia smooth; on ferns .... GRIGGSIA
            - (b). Perithecia with setae or septate hairs..... CIFERRIOMYCES
            - (c). Perithecia with a stromatic, radiate, fissured crown....... .... ERIKSSONIA
          - (2). Stroma strongly erumpent from a hypostroma; with several perithecia
            - (a). Asci 8-spored .... COCCOSTROMA
            - (b). Asci 1-4 spored .... ZIMMERMANNIELLA

II. Ascospores 2-celled
  A. Ascospores yellow-brown to brown
    1. Ascospores with a basal hyaline small cell. . . . . COCCOCHORELLA
    2. Ascospores with a central septum . . . . GIBELLINA
  B. Ascospores hyaline to yellowish
    1. Ascospores unequally 2-celled (apiosporous)
      a. Stromata subcuticular, hemispherical . . . . REHMIODOTHIS
      b. Stromata subepidermal, elongate; septum of ascospore below the middle . . . . APOSPHAERIA
      c. Stroma subepidermal; septum of the ascospore above the middle . . . . STIGMOCHORA
    2. Ascospores equally 2-celled
      a. Stromata immersed
        (I). Clypeus present; on legumes . . . . STIGMOCHORA
        (II). Clypeus blackening limited; on *Prunus*. . . . . POLYSTIGMELLA
      b. Stromata erumpent, pulvinate . . . . PHYLLOCREA
III. Ascospores filiform. . . . . OPHIODOTHIS

Physalospora: Some three hundred species with single immersed perithecia and one-celled hyaline ascospores have been described under this generic name. Von Arx and Müller (**1954**) have limited the genus to those mostly parasitic species with rather large thick-walled ascospores and thin-walled unitunicate asci, which they placed in the Polystigmataceae. Most previously described species were transferred by these authors to a variety of other genera. Those with bitunicate asci are related to *Botryosphaeria.*

The perithecia in this genus are borne singly beneath the epidermis and occur mostly as parasites on the leaves of various plants. Setae may occur about the ostiole of some species (*Acanthorhynchus, Trichophysalospora*). The ascospores are rather large and have a milky or granular pale yellow to olive-green content.

Plectosphaera: The perithecia in this genus occur singly within the leaf tissue, or in some species with a slight development of stromatic hyphae, in which case they are transitional to the genus *Phyllachora*. It also differs from *Physalospora* in the smaller hyaline ascospores without any granular content. The light colored perithecial wall relate it to *Hyponectria* and the Amphisphaeriaceae. It is also similar to *Glomerella* and the Diaporthaceae. All these families intergrade.

Phyllachora: This is another large genus with some six hundred described species. The type species is *P. graminis*, a name used for the species on grasses. Orton (**1944**) has recognized some forty-six species on genera of grasses. Many other species occur on the leaves of tropical plants, and have been distinguished by their host occurrence. As a result many names have become synonyms. The genus was originally placed in the Dothideales. Theissen and Sydow (**1915**) placed it in their family

Phyllachoraceae, in this order, along with other genera having a clypeus. The works of Petrak (**1924**), Orton (**1924**) and Miller (**1951, 1954**) have shown that the development of the perithecium is of the ascohymenial type. Petrak (**1924**) suggested its relation to the Hypocreales through such genera as *Epichloe* and *Ophiodothis*. The perithecia are imbedded in a more or less well developed stroma beneath a blackening of the surface host tissue, the clypeus. The asci are unitunicate; there are true paraphyses and the ascospores are ellipsoid, one-celled and hyaline.

Orton (**1924**) has studied the development of *P. graminis* and Miller has reported on that of *P. ambrosiae* and *P. lespedezeae.*

In *P. graminis*, infection occurs in the spring from ascospores of overwintered stromata. The perithecia arise as coils of fine hyphae beneath the epidermis within the stromatic area. Paraphyses grow upward from the basal portions of the perithecial wall and fill the interior of the perithecium. The asci grow up from the base between these filaments. The upward growth of hyphae from the upper wall of the perithecium penetrates into the upper epidermis causing a blackening of this tissue and the formation of a clypeus. Other hyphae grow downward into the lower epidermis forming a clypeus in this tissue. No compact stroma is formed.

In *P. ambrosiae*, infection causes black spots in both surfaces of the leaf. A loose stroma of mycelial strands is formed within the leaf. Spermagonia appear as conic locules just beneath the epidermis. The spermatia in these pycnidia are minute, fusoid, one-celled and hyaline. The perithecia arise in the stromatic area as coils of fine hyphae. Hyphae from these coils were seen in contact with spermatia but no actual fusion was observed. The central hyphae of this coil arrange themselves in a radiate manner from the base and sides toward the ostiolar end of the perithecium. Some of these hyphae give rise to croziers with paired nuclei and subsequently to asci. Other sterile hyphae grow upward between the asci as true paraphyses. An ostiole is formed and penetrates through the clypeus. The ascospores are ejected through a pore in the apical wall.

Primary infections of *P. lespedezeae* give rise to yellow spots containing spermagonia and scattered perithecia beneath individual clypei. The development of the fruit body is similar to that of *P. ambrosiae*. The spermatia are small and ellipsoid. No fusion with fungus hyphae was seen. In late summer larger stromata originate within the leaf tissue. These stromata have a blackened outer cortex and a hyaline pseudoparenchymatic core. They usually remain sterile, but occasionally they may contain perithecia which mature the following spring. Ascospores from these perithecia cause the primary infections of the following season.

Polystigma: This genus was originally placed in the Hypocreales because of the fleshy, yellow or reddish color of the stromata. The perithecia which are imbedded in this stroma have filiform paraphyses growing upward from the base and not downward as in the case with the pseudoparaphyses characteristic of the Hypocreales as distinguished by Luttrell. The asci are long clavate, with a long stalk and rounded at the apex. The ascospores are one-celled, ellipsoid and hyaline.

Polystigma rubrum is the cause of a red spot disease of cherry leaves in Europe and Asia. Perithecial development has been studied in this species by Fisch (**1882**), Frank (**1883**), Blackman and Welsford (**1912**) and Nienburg (**1914**). Infection takes place in the early spring. The entire thickness of the leaf is penetrated by a mycelial stroma which is fleshy and yellow-orange in color. Some of these hyphae can be seen penetrating through the stomata, and Frank and Fisch considered these as trichogynes. Spermagonia arise just beneath the epidermis as cylindric cavities within the stroma. They contain numerous filiform, hamate, hyaline spermatia. Frank and Fisch claimed that there was fusion between these spermatia and the so-called trichogynes. Blackman and Welsford and Nienburg failed to see any such fusion.

The perithecia arise in the stroma in early summer as a coiled archicarp of larger hyphae. According to Nienburg, this coil consists of a multinucleate basal cell, a multinucleate second cell, two following uninucleate cells and a number of smaller apical cells which he considers a trichogyne. Such trichogynes do not usually extend above the surface of the leaf. He supposed that the second multinucleate cell acted as an antheridium and passed a nucleus through a pore into the third uninucleate cell, which acted as an ascogonium. The two nuclei in this female cell pair but do not fuse. Blackman and Welsford believed that the ascogonium disintegrated and that the ascogenous hyphae arose independently from the basal wall hyphae of the perithecium. Nienburg states that all of the archicarp cells, except the ascogonial cell, degenerate but that the ascogenous hyphae arise from this binucleate sexual cell. As a result of nuclear pairing the surrounding vegetative cells are stimulated to elongate in a radial fashion toward the ostiolar end of the perithecium to form the true paraphyses of the mature ascocarp.

The spermagonial state of *Polystigma* has been placed in the form genus *Polystigmina* with hamate conidia. Species of the genus *Physalospora*, which have their stromata on genera of the Leguminosae and have a spermagonial state in the form genus *Rhodosticta*, with cylindric spermatia, are considered as synonymous with *Polystigma*.

Uropolystigma: This is a small genus similar to *Polystigma* in causing parasitic hypertrophies on leaves. The perithecia are imbedded in the under side of a fleshy light colored stroma with a blackened clypeus on the upper surface. It differs in the ascospores which are one-celled and hyaline but with a tapered elongate upper end.

Schizochora: This genus is similar to *Phyllachora*, with a blackened clypeus, but with ascospores having a terminal appendage. It is a doubtful genus.

Diachora: Diachora forms a light colored stroma filling the entire thickness of the host leaf and staining blue in KI. There is a blackened clypeus on both surfaces of the leaf. The asci arise from the side walls of the perithecium, the base remaining sterile. The ascospores are ellipsoid, one-celled and hyaline. A conidial state, *Diachorella*, occurs on the living leaves. It forms a hymenial layer of conidiophores, from which pyriform conidia are abstricted, in the blackened clypeate layer of the stroma.

Griggsia: The perithecia of *Griggsia* are superficial upon an internal mycelial stroma. The asci are broad clavate and the ascospores are large, ellipsoid, one-celled and hyaline.

Ciferriomyces: This genus has superficial perithecia borne singly upon an internal hypostroma, but the ascocarp is covered with spines or septate hyphae. The ascospores are one-celled and hyaline.

Erikssonia: In this genus the perithecia are clustered in linear series within the leaf tissue, but are erumpent as a crown-like stroma which is radiately sulcate. The ascospores are one-celled, hyaline or pale yellow.

Coccostroma: This genus includes all species with erumpent pulvinate stromata on a basal hypostroma and containing several perithecia. The stromata are pale in color, or with a blackened cortex and may have a broad or stipe-like base. The ascospores are one-celled, hyaline at first, but may become olive-brown at maturity. Von Arx and Müller include *Bagnisiopsis* in this genus.

This genus has been placed in both the Sphaeriales and the Dothideales. Shirakawa (**1956**) has studied the development of *C.* (*Bagnisiopsis*) *nuda* and *C. polymorpha*. These species formed disc-like stromata on the under side of leaves of *Cibotium*, a tree fern. These stromata were composed of hyaline thin-walled parenchyma and had a cortex of thick-walled brown cells. The perithecia arose within the hyaline parenchyma as light colored ovoid areas. Dark staining cells could be seen in these areas and they were assumed to be ascogonia. Dark staining hyphal strands appeared in these areas and grew upwards, filling the cavity as the parenchyma disintegrated. These hyphae formed a loosely interwoven perithecial wall. The paraphyses disintegrated and ascogenous cells appeared in the base of the cavity, from which the asci arose. This formation of a differentiated area within a stroma and the indefinite archicarp formation is characteristic of the Ascoloculares. The asci on the other hand are given as being thin-walled and unitunicate and with an indefinite apical pore. The perithecial wall is indefinite and formed as a differentiation of the wall tissue. Similar differentiated areas of wall tissue are to be found in the Dothideales. The paraphyses are of an indefinite type. This is one of those types of development which allows different interpretations.

Zimmermanniella: This genus also has erumpent stromata from a basal hypostroma, with many perithecia, as in *Coccostroma*, but the asci have one to four, usually three ascospores.

Rehmiodothis: In *Rehmiodothis* the stromata are subcuticular and hemispheric, with a flattened base and contain one or several perithecia. The surface cuticle is blackened to form a clypeus. The ascospores are ellipsoid, two-celled, with a minute lower cell, and hyaline to yellowish.

Apiospheria: This genus has a light colored stroma in leaves, with a blackened clypeus above each perithecium. The ascospores are two-celled, apiosporous, with a minute lower cell and hyaline to yellowish. A conidial state of the form genus *Rhodosticta* has been reported. It consists of pycnidia sunken in the stroma which contain filiform, hamate, curved hyaline conidia.

Stigmochora: This genus has light colored stromata containing several perithecia, in living leaves. There is a blackened clypeus in the epidermis. The ascospores of this genus are equally two-celled, or with the upper cell smaller than the lower, and hyaline or yellowish.

Polystigmella: The light colored stromata of this genus contain pycnidial locules in the living leaves of *Prunus*. This *Polystigmina* state has filiform hyaline conidia. The perithecia occur in the overwintered leaves and the ascospores are two-celled and hyaline, in which character this genus differs from *Polystigma*.

Coccochorella: The perithecia are formed beneath a blackened clypeus in living leaves. The ascospores are two-celled and brown with a small hyaline lower cell, otherwise as in *Sphaerodothis*.

Gibellina: *Gibellina* has light colored stromata containing seriate perithecia on stems of grasses. The ascospores are light brown and equally two-celled. *Gibellina cerealis* is the cause of a disease of cereal grains (Cavara, **1893**).

Isothea: This is a monotypic genus (*I. rhytismoides*) on alpine plants of *Dryas*. It forms reddish spots on the leaves, containing several perithecia within a stromatic area. The ostioles do not break through the superficial clypeus of blackened surface tissue. The ascospores are one-celled and hyaline.

Phylleutypa: Phylleutypa occurs on stems as an effuse stroma with an erumpent blackened surface. The ascospores are one-celled and hyaline.

Lohwagia: This genus causes gall-like swollen erumpent stromata with a roughened blackened surface and elongate sulcate ostioles. The ascospores are one-celled and hyaline to yellow-brown.

Sphaerodothis: Sphaerodothis has dark colored, brittle, carbonaceous stromata. A blackened clypeus may be formed on one or both surfaces of the host leaf. The asci disintegrate early in the perithecium. The ascospores are ellipsoid to globose and brown in color.

Phaeochorella: This genus forms a dark colored parenchymatous, subepidermal stroma with a blackened surface clypeus. The ascospores are one-celled, dark brown and with a hyaline band about the equator. A conidial state, *Phomachorella,* is known which forms a layer of conidiophores in a stroma identical to the ascus stroma, from which ellipsoid, one-celled, hyaline conidia are abstricted in a gelatinous mass.

Pseudotthiella: The stroma in *Pseudotthiella* is disc-like, black and strongly erumpent-superficial. The ascospores are one-celled, somewhat flattened, olive brown and with a hyaline band about the equator. A conidial state, *Pseudothiopsella*, occurs in the same stromata as pycnidial locules containing long fusoid, flattened, olive brown, conidia with a similar hyaline equatorial band.

Phyllocrea: This genus has light brown stromata containing several perithecia. The stromata are pulvinate and erumpent-superficial. The genus was described in the Hypocreaceae. It differs from *Coccostroma* in the two-celled, hyaline ascospores. A conidial state occurs in the surface of the stroma as a hymenium of conidiophores bearing clavate one- or two-celled conidia.

AMPHISPHAERIACEAE:

Winter established the family Amphisphaeriaceae for genera with superficial perithecia, similar to those of the Melanommeae, but with the base partially immersed in the substratum. Müller and von Arx (**1962**), based the family on the type genus *Amphisphaeria*. The basis was the presence of a small plug or disc, which stains blue with potassium iodide, in the thickened apex of the ascus. As a result genera from a number of previously described families have been transferred to the Amphisphaeriaceae. It included the genera placed by Höhnel (**1919**) in his Physosporelleen and by Petrak (**1924**) in his Hyponectriaceae, and also genera from other families. The amerosporous members of the family were previously placed by von Arx and Müller (**1954**) in their family

Polystigmataceae. The presence of the blue staining plugs in the ascus tip and the germ pores in the ascospores of some genera also relate these genera to the Xylariaceae. For the most part, the family includes genera with simple immersed perithecia with little or no stromatic development. The ascospores are various and there are probably other genera with asci staining blue with KI which will eventually be placed in this family.

I. Ascospores 1-celled
  A. Perithecia lying horizontally, in wood, beneath a clypeus .... XYLOCHORA
  B. Not as above
    1. Perithecia in valsoid groups; ostioles collectively erumpent... MYELOSPERMA
    2. Perithecia immersed singly
      a. Perithecial wall with light colored cells .... HYPONECTRIA
      b. Perithecial wall with dark colored cells
        (I). Ascospores appendaged; asci sessile................ UROSPORELLA
        (II). Ascospores not appendaged; asci stalked........... ANISOSTOMULA
          (Perithecia larger; ostioles more definite).......... PHYSOSPORELLA
II. Ascospores 2-celled
  A. Ascospores unequally 2-celled (apiosporous)
    1. Ascospores appendaged, brown, with germ pore in the larger cell ....
    .... APIORHYNCHOSTOMA
    2. Ascospores not appendaged, hyaline to brown, without germ pores
      a. Perithecia with setae about the ostiole; ascospores yellow to brown ......
      .... CHAETAPIOSPORA
      b. Perithecia glabrous; ascospores hyaline
        (I). Perithecia lying horizontal; ostiole lateral............ APIOTHYRIUM
        (II). Perithecia vertical; ostiole apical
          (A). Perithecia seriate in a linear stroma .... APIOSPORA
          (B). Perithecia singly immersed in the host........ PSEUDOMASSARIA
  B. Ascospores equally 2-celled
    1. Ascospores appendaged
      a. Perithecia hemispheric with a flattened base, ascospores brown.. SEYNESIA
      b. Perithecia globose to ellipsoid
        (I). Spores fusoid, ends tapered
          (A). Perithecia flattened ellipsoid; beneath a clypeus
            (1). Perithecia horizontal; ostioles lateral .... OXYDOTHIS
            (2). Perithecia vertical; ostioles apical......... MERRILLIOPELTIS
          (B). Perithecia globose, without a clypeus .... CERIOSPORA
        (II). Spores ellipsoid, ends rounded .... CERIOPHORA
    2. Ascospores not appendaged
      a. Perithecia hemispheric, with flattened base
        (I). Spores brown, ellipsoid, striate .... ROUSSOELLA
        (II). Spores hyaline, smooth .... LEIOSPHAERELLA
      b. Perithecia globose to ellipsoid; ascospores ellipsoid
        (I). Ascospores with germ pores in each end
          (A). Ascospores hyaline, becoming brown; without pseudostroma or clypeus .... CAINIELLA
          (B). Ascospores borne with 8 vertical striae and a gelatinous envelope; pseudostroma or clypeus present........................ CAINIA
        (II). Ascospores without germ pores
          (A). Perithecia in diatrypoid stroma.......... GRIPHOSPHAERIOMA
          (B). Perithecia not in diatrypoid stroma
            (1). Ascospores brown .... AMPHISPHAERIA

(2). Ascospores hyaline to yellowish .... LEIOSPHAERELLA
C. Ascospores more than 2-celled
 1. Ascospores brown
  a. Perithecia scattered singly.......................... HYMENOPLELLA
  b. Perithecia in an effuse stroma .... LEPTEUTYPA
 2. Ascospores hyaline
  a. With a superficial blackened clypeus
   (I). Perithecia singly, in series........................ MONOGRAPHELLA
   (II). Perithecia beneath a clypeus.................... CLYPEOTHECIUM
   (III). Perithecia in an elongate stroma .... EXARMIDIUM
  b. Without a clypeus
   (I). Perithecia immersed, single or clustered
    (A). Spores ellipsoid....................... GRIPHOSPHAERELLA
    (B). Spores fusoid.............................. MONOGRAPHUS
   (II). Perithecia in diatrypoid stroma .... GRIPHOSPHAERIOMA
D. Ascospores muriform, hyaline
 1. Stroma valsoid .... CLATHRIDIUM
 2. Perithecia scattered or clustered .... GRIPHOSPHAERIA

Xylochora: This is a genus erected by Müller and von Arx for a species (*X. craticola*) occurring on dead wood. It has perithecia lying horizontally in the substrate and with a laterally erumpent ostiole, which is surrounded by a brownish hyphal stroma. The ascus tip stains blue with KI. The ascospores are ellipsoid, one-celled and hyaline.

Myelosperma: The perithecia in this genus are grouped in a valsoid manner within the host tissue and the ostiolar necks are collectively erumpent. In these respects it is similar to *Cryptosporella*. The asci, however, are long cylindric and lie parallel in a hymenial layer. The ascospores are uniseriate, ellipsoid, one-celled, hyaline, and have a thin gelatinous envelope.

Hyponectria: This genus forms perithecia on dead leaves of *Buxus*. The perithecial wall is light in color except for a slight blackening about the ostiole. The ostiole is poorly developed. The asci are clavate with a definite stalk and have an apical plug which stains blue with KI. The ascospores are one-celled and hyaline. This genus and *Anisostomula* were placed by Petrak in his family Hyponectriaceae.

Anisostomula: This genus is similar to *Hyponectria* but has smaller perithecia and darker colored perithecial walls and is found on leaves of *Quercus*.

Physosporella: This is given by von Höhnel (**1919**) as differing from *Anisostomula* in the larger perithecia and more definite ostiole. Von Arx and Müller (**1954**) give it as a synonym of *Plectosphaera* which they place in their Polystigmataceae because the asci have no blue staining tip with KI.

Urosporella: This genus has simple immersed perithecia and fusoid, one-celled, hyaline ascospores with a filamentous appendage at each end of the spore. The asci show the blue stain with KI.

Oxydothis: This genus has several flattened ellipsoid perithecia lying horizontally in a superficial clypeus-like stroma which forms spots on the stems of grasses, palms, etc. The ostioles are lateral and bent at an angle of 90° to penetrate the epidermis. The asci have a disc in the apex which stains blue with KI. The ascospores are long fusoid, two-celled, hyaline and have a filiform appendage at each end.

Merrilliopeltis: This genus is given by Müller and von Arx (**1962**) as a synonym of *Oxydothis*. It differs in having a single perithecium which is vertical in the substratum with an apical ostiole erumpent through a clypeus.

Ceriospora: This genus has globose immersed perithecia with dark colored walls and two-celled, fusoid hyaline ascospores with a filiform appendage at each end. The asci have plugs, which stain blue with KI, in the apex.

Ceriophora: This genus has similar globose immersed perithecia and asci with blue staining plugs with KI, but the ascospores are hyaline to light brown and have more rounded ends, are striate and have a tapered appendage at each end.

Seynesia: This genus has single hemispherical perithecia with a flattened base beneath a blackened clypeus. The asci have the blue staining disc in the apex. The ascospores are two-celled, brown and have a gelatinous sheath and apical appendages at each end.

Pemphidium: This genus is given by von Arx and Müller (**1954**) and by Müller and von Arx (**1962**) as probably an immature *Seynesia*. The ascospores were given as one-celled and hyaline, but they may become two-celled.

Roussoella: This genus forms similar black hemispherical subepidermal stromata which contain several perithecia with light colored walls, on bamboo. The asci are cylindric and have apical apparatus which stains blue with KI. The ascospores are two-celled, brown and have longitudinal striae.

Leiosphaerella: In this genus the perithecia are immersed in stems or leaves. They may be globose or with a flattened base. The ascospores are two-celled and hyaline.

Amphisphaeria: There has been some difference of opinion as to the type of this genus. Müller and von Arx give *A. umbrina* as the type. The genus includes thick-walled perithecia entirely or partially sunken in the woody substrata. The asci have spots in the thickened apex which stain blue with KI. The ascospores are ellipsoid to cylindric, two-celled and yellow-green to dark brown. *Phorcys*, with gelatinous envelopes on the spores, is given as a synonym.

Apiospora: This genus forms elongate dark colored stromata in the stems of grasses. The stromata contain several perithecia in a linear series, and rupture by a longitudinal slit. The ascospores are two-celled, hyaline and apiosporous, with a larger upper and a smaller ventral cell. Theissen and Sydow placed the genus in the Dothideales, but the asci are unitunicate with a blue stain in the apical wall.

Pseudomassaria: This genus has large perithecia immersed in leaves or stems, singly without any stromatic development. The asci are clavate, somewhat stalked and with an apical apparatus staining blue with KI. The ascospores are two-celled, hyaline and with the septum toward the lower portion of the spore.

Apiothyrium: This genus has its perithecia lying horizontally between the cuticle and the epidermis, with a laterally placed ostiole. There is a slight development of stroma about the perithecium. The asci do not stain blue in KI. The ascospores are two-celled, hyaline and apiosporous.

Apiorhynchostoma: This genus has globose immersed scattered perithecia with some hyphal outgrowth about the ostiole. The asci do not stain blue. The ascospores are somewhat unequally two-celled, and brown. The upper larger cell has a small hyaline appendage and the lower smaller cell has a terminal germ pore.

Chaetapiospora: This genus has sunken scattered perithecia with dark colored setae about the ostiole. The asci have an apical apparatus staining blue with KI. The ascospores are unequally two-celled and yellowish to brown.

Cainia: This genus has large sunken perithecia surrounded by a development of dark colored hyphae. The asci are clavate with a complex apical apparatus staining blue with KI. The ascospores are two-celled and brown and surrounded by a gelatinous envelope. The ascospores are octagonal in cross section and have eight vertical striae corresponding to the angles of the spore. There is a germ pore at each end of the spore.

Cainiella: This genus has globose perithecia with an erumpent ostiole. The asci have a thickened apical wall through which there runs a canal surrounded by a collar which stains blue with KI. The ascospores are two-celled, asymmetric, stout ellipsoid, bent at the septum and hyaline at first, becoming pale brown.

Hymenoplella: This genus has superficial perithecia and olive-yellow, four-celled ascospores.

Lepteutypa: The perithecia of *Lepteutypa* develop in a subhyaline loosely woven mycelium in the substrate. The ascospores are colored, thick-walled and several septate.

Exarmidium: This genus has a linear stroma with a blackened surface similar to the clypei of the Phyllachoraceae. The ascospores are ellipsoid, hyaline and three-septate. *Clypeothecium* differs chiefly in the presence of a small clypeus above each perithecium.

Monographella: This genus has small globose, scattered subepidermal perithecia with a hyaline to light colored wall. The ascospores are fusoid, many-celled and hyaline.

Monographus: This genus has black, stromatic, subepidermal perithecia and fusoid hyaline ascospores with three to six guttulae, resembling septa.

Griphosphaerella: This genus has immersed, scattered, or clustered perithecia immersed beneath the epidermis. The ascospores are fusoid-cylindric and three to five septate and hyaline.

Griphosphaeria: This genus has similar immersed and clustered perithecia. The asci are long cylindric and the ascospores are uniseriate, ellipsoid, hyaline and usually three-septate, but sometimes with a vertical wall in one of the cells.

Griphosphaerioma: This genus is similar to *Griphosphaeria,* but the perithecia are in a diatrypoid stroma and the ascospores are mostly two-celled (sometimes four-celled).

Clathridium: This genus was originally described as having muriform, hyaline, ascospores. Its type (*C. burchellii*), according to Berlese, was a *Diaporthe.* He gave *C. massarinum* with three or more transverse septa as the type species. Müller and Loeffler (**1957**) give *Discostroma, Phragmodothella, Curreyella* and *Griphospherioma* as synonyms. The ascospores of these species have two or more transverse septa and occasionally vertical septa in some cells. The spores are hyaline. The stromata are immersed or erumpent with several perithecia and have a light colored central tissue and a blackened clypeus or cortex. *Griphospherioma* was described as having a diatrypoid stroma. The asci are cylindric with an apical ring which stains blue with KI.

SPHAERIACEAE:

This is a purely artificial family used to include a number of genera which have not been placed or do not belong in any of the previously described families. The genera included are, for the most part, saprophytes with superficial perithecia. The asci have simple thickened apical walls with a disc-like thickening not staining with KI. They are found in the older systems in the families Trichosphaeriaceae, Melanommeae and Amphisphaeriaceae.

The following key includes some of these genera:

- I. Ascospores 1-celled .... TRICHOSPHAERIA
- II. Ascospores 2-celled
  - A. Ascospores separating into separate cells within the ascus . TRICHOSPHAERELLA
  - B. Ascospores not separating into cells
    - 1. Perithecia with an elongated ostiolar neck
      - a. Ascospores hyaline .... KLASTERSKYA
      - b. Ascospores brown .... RHYNCHOMELIOLA
    - 2. Perithecia with a short neck or punctiform ostiole
      - a. Perithecia with setae or hyphal bundles
        - (I). Perithecia with elongate hyphal bundles about the ostiole .... PSEUDORHYNCHIA
        - (II). Perithecia with simple setae
          - (A). Hyperparasites with thin-walled often slimy perithecia ..... RIZALIA
          - (B). Perithecia thick-walled, not slimy
            - (1). Perithecia upon a hyphal subiculum .... NIESSLIA
            - (2). Perithecia upon a parenchymatic stromatic crust .... ERIOSPHAERIA
      - b. Perithecia glabrous
        - (I). Perithecia rugose, mammillate on surface .... BERTIA
        - (II). Perithecia not rugose
          - (A). Perithecia in erumpent stalk-like stromata .... REHMIOMYCELLA
          - (B). Perithecia not in erumpent stromata
            - (1). Basal stroma light in color; spores large (20 μ or more) .... MELCHIORIA
            - (2). Basal stroma slight or dark in color; saprophytes .... CHAETOSPHAERIA
            - (3). Hyperparasites, immersed in a host stroma.. RHAGADOSTOMA
- III. Ascospores many-celled
  - A. Ascospores hyaline
    - 1. Perithecia setose or hairy
      - a. With brown setae .... ACANTHOSTIGMA
      - b. With yellow-green hairs when young .... PSEUDOTRICHIA
    - 2. Perithecia glabrous, beaked .... CERATOSPHAERIA
  - B. Ascospores brown
    - 1. Perithecia with a clypeus .... CLYPEOSPHAERIA
    - 2. Perithecia without a clypeus
      - a. Ascospores fragmenting .... OHLERIA
      - b. Ascospores not fragmenting; appendaged .... REBENTISCHIA
      - c. Ascospores with central cells brown and terminal cells hyaline BROOMELLA
- IV. Ascospores muriform
  - A. Ascospores hyaline; perithecia beaked .... RHAMPHORIA
  - B. Ascospores brown, perithecia rugose, mammilate .... CROTONOCARPIA
- V. Ascospores filiform, appendaged, septate .... DILOPHIA

Trichosphaeria: This genus has superficial perithecia covered with stiff setae or hairs. The species occur on decaying wood and their asci have eight one-celled, hyaline ascospores.

Trichosphaerella: This genus has similar superficial perithecia, often on a slight subiculum on decaying wood. The asci are cylindric and contain eight two-celled hyaline ascospores in which the cells separate to form sixteen part spores in a single row.

Niesslia: This genus has superficial setose perithecia and elongate fusoid to cylindric two-celled hyaline ascospores which do not fragment.

Eriosphaeria: This genus also has superficial, setose perithecia, but in this genus they are seated on a thin parenchymatic subiculum or hypostroma. The ascospores are two-celled and hyaline. Species with sixteen ascospores in the ascus have been placed in the genus *Eriosphaerella.*

Chaetosphaeria: In *Chaetosphaeria* the perithecia are glabrous and superficial upon a subiculum or thin hypostroma on the surface of the substrate. The ascospores are one- to three-septate and hyaline to colored depending upon maturity. Conidial states consisting of brown, septate conidiophores arising from the mycelial subiculum are present. In some species these belong to the form genus *Catenularia* or *Sporoschisma* (Hughes, 1965) in which the conidia are angular or cylindric, brown, and formed endogenously from terminal phialides on simple conidiophores. In other species the conidial state belongs to the form genus *Menispora*, and has one-celled, curved or falcate conidia arising from phialides on branched conidiophores. The genus has been monographed by Booth (1957, 1958). *Chaetosphaeria* (*Melanopsamma*) *pomiformis* has a *Stachybotrys* conidial state which bears greenish-yellow, one-celled conidia, which are black in mass, from clusters of seven to ten phialides borne on the bulbous apex of the conidiophore.

Rizalia: The perithecia of this genus are hyperparasites on other leaf fungi. The perithecia have thin slimy walls and may be setose. They are seated on a mycelial subiculum which may also have setae. The ascospores are fusoid to elongate, two-celled and hyaline.

Melchioria: The perithecia in this genus are borne singly or clustered in a superficial light colored basal stroma on the surface of the host. They may have a sparse tomentum. The ascospores are fusoid, two-celled, hyaline and may have a gelatinous envelope.

Pseudorhynchia: This monotypic genus has superficial perithecia upon a basal mycelium. There are elongate bundles of hyphae growing out from the wall about the ostiole. The genus was placed in the Chaetomiaceae by von Höhnel, but the asci are persistent and the ascospores are one- to two-celled and hyaline.

Klasterskya: The perithecia in this genus are also superficial upon a basal mycelium, but the ostiolar necks are long and filiform, and are penicillate at the apex. The ascospores are cylindric, two-celled and hyaline.

Rhynchomeliola: This genus also has perithecia with long filiform ostiolar necks upon a basal mycelium from which setae may arise. The ascospores are two-celled, colored and often echinulate.

Rhagadostoma: The species of this genus are hyperparasites upon lichens and other Ascomycetes. The perithecia are immersed within a stroma on the host. The ascospores are two-celled and hyaline or slightly colored.

Rehmiomycella: This genus has a basal foot or hypostroma with an erumpent stalk bearing one or more perithecia. The ascospores are two-celled and hyaline.

Bertia: *Bertia* has superficial perithecia clustered upon dead wood, with carbonaceous and thick rugose, mammillate walls. The asci are stipitate and the ascospores are spindle-shaped, two-celled and hyaline.

Acanthostigma: This is a genus with superficial, setose perithecia and hyaline many-celled ascospores.

Pseudotrichia: This genus is similar to *Acanthostigma*. The perithecia are found on dead wood or the stromata of other Ascomycetes and their surface is covered by a felt of yellow-green tomentum when young. Wehmeyer (**1941**) reported that *P. aurata* has downward-growing pseudoparaphyses in the perithecium, as in the Hypocreales. The ascospores are two- to four-celled, fusoid and hyaline.

Ceratosphaeria: This genus has immersed perithecia with elongate ostioles and hyaline, many-celled ascospores.

Clypeosphaeria: This genus has perithecia which are erumpent through a blackened clypeus. The ascospores are many-celled and brown.

Ohleria: In *Ohleria* the perithecia are superficial upon a thin stroma. The ascospores are four-celled, brown and fragment into two two-celled part spores.

Rebentischia: In this genus the perithecia are erumpent and the ascospores are four-celled, brown and with a basal hyaline appendage.

Rhamphoria: This genus has perithecia which are erumpent with an elongate ostiolar neck and ascospores which are muriform, hyaline and have a gelatinous envelope.

Crotonocarpia: This genus has perithecia which are superficial with a flattened base and a mammillate outer wall as in *Bertia*, but with ascospores which are muriform and brown.

Dilophia: The immersed perithecia of this genus have ascospores which are filiform, septate, hyaline and with a filiform appendage at each end. *Dilophia graminis* is a leaf parasite on grasses. The conidial state is *Diplophospora graminis*, which occurs as a pycnidium within a blackened stroma-like spot upon the leaves.

Broomella: This genus has perithecia which are immersed or erumpent and borne singly or fused into a compound stroma. The asci are thin-walled and have an apical cap which does not stain blue with KI. The ascospores are four-celled. The two central cells are larger and dark brown and the two end cells smaller and hyaline with a filiform appendage at each end.

Shoemaker and Müller (**1963**) recognized four species which are very similar in their ascus state but with differing conidial states. All conidia are four-celled with brown central cells and hyaline end cells with terminal appendages, as in the ascospores, but borne in different types of fruiting bodies. In *B. vitalbae* and *B. acuta* the conidia are borne in widely open acervuli as in the Melanconiaceae. The conidia have a single apical setula (*Monochaetia*), in the former, and more than one apical appendage (*Pestalotia*), in the latter species. In *B. montaniensis*, the conidia are borne in a synnema (*Arthrobotryum*) as in the Stilbaceae. In *B. excelsa*, the conidia are in a more enclosed pycnidial structure (*Ahmadinula*). These authors consider all these forms as belonging to the genus *Pestalotia*.

Broomella is similar to *Ceriospora*, but the asci do not have the apical ring staining with KI, as in that genus and the central cells in the ascospores are brown in *Broomella*. *Keissleria* is a synonym.

HALOSPHAERIACEAE:

The Halosphaeriaceae was used as a family name by Müller and von Arx (**1962**) for a group of genera of marine habitat which were not similar to any previously described genera. A large number of species have been described growing in sea water on submerged or drift-wood, piling, cordage, etc. as well as on marine algae and higher plants. Many of these species have been placed in previously described genera. The newly described genera considered here show relationships to many diverse genera, families and orders, but differ in several common characteristics. Nearly all the species have some type of appendages on the ascospores. These appendages serve as floats in the distribution of the spores and function in the attachment to a favorable substratum in moving water. Many of the asci are deliquescent and release the ascospores within the perithecial centrum. The ascocarps are superficial or partially immersed.

Very few reports on the development of the ascocarp are available, but both the ascostromatic and the ascohymenial type of development have been mentioned and both unitunicate and bitunicate asci occur. There are many variations in the formation of the interthecial tissues within the family, which do not fit clearly into the Luttrell classification. As a result there has been some difference of opinion as to the use of the development of the perithecial centrum or the type of ascospore appendages as a basis for generic differentiation, and generic limits are in some cases vague or have been altered. Until more information is forthcoming, it is probably more practical to keep these genera together in this family.

I. Perithecia light colored, fleshy, ostiolate
  A. Spores fusiform, curved, 4-celled .... ORCADIA
  B. Spores with a long tapered filiform base, 1-3-celled .... TRAILIA
II. Perithecia dark colored
  A. Ascospores with appendages
    1. Ascospores filiform
      a. Ascospores 1-celled, apical appendages tapered.............. LULWORTHIA
      b. Ascospores septate, apical appendages rounded................. LINDRA
    2. Ascospores not filiform
      a. Ascospores with the wall merely thickened at the ends
        (I). Asci deliquescing and releasing immature ascospores into the perithecial cavity................................. CHADEFAUDIA
        (II). Ascospores released after maturity .... HALOGUIGNARDIA
      b. Ascospores with apical gelatinous caps, 1-3-celled ..... LENTESCOSPORA
      c. Ascospores with a disc-shaped lateral appendage at the septum......... .... PARALIOMYCES
      d. Ascospores with fine, spine-like appendages
        (I). Ascospores 1-celled; appendages all over the surface; ascocarp a cleistothecium................................. PLECTOLITUS
        (II). Ascospores 2-celled; appendages numerous at each end .... HERPOTRICHIELLA
        (III). Ascospores 2-celled with a single long filiform appendage. LORAMYCES
        (IV). Ascospores 2- to 7-celled; appendages both terminal and equatorial .... PERITRICHOSPORA
      e. Ascospores with an irregular gelatinous envelope or appendage
        (I). Ascospores 1-celled, with an undulate sheath .... SAMAROSPORELLA
        (II). Ascospores 2-celled, with an undulate envelope .... DIDYMOSAMAROSPORA
      f. Ascospores with firm, band-like appendages
        (I). Ascospores 2-celled
          (A). Ascospores with a single appendage at each end and with or without appendages at the septum................ CERIOSPOROPSIS
          (B). Ascospores with 2 appendages at each end at right angles to one another .... ANTENNOSPORA
          (C). Ascospores with 2-9 appendages at each end .... REMISPORA
          (D). Ascospores with an apical appendage at each end and with a curved or irregular envelope at the septum .... HALOSPHAERIA
          (E). Ascospores with 3 curved appendages at each end .... ARENARIOMYCES

(II). Ascospores with more than 2 cells
(A). 3-4 radiating appendages at 1 or both ends ...... TORPEDOSPORA
B. Ascospores without appendage
1. Ascospores fusoid to ellipsoid. 2-celled .... LIGNICOLA
2. Ascospores long fusoid to filiform, 1-celled .... MAIREOMYCES

Loramyces: Loramyces occurs on the stems of Cyperaceae, Juncaceae and *Equisetum*, in salt water. The perithecia are light colored, rather large and superficial. The ascospores are two-celled and hyaline and with a long filiform appendage at one end and a gelatinous envelope. Müller and von Arx (**1962**) place it in the Sphaeriaceae.

Remispora: The perithecia of this genus are light colored, of loosely arranged thin-walled cells. The asci arise from a basal dome-shaped parenchyma. The ascospores are two-celled with two to several gelatinous or stiff appendages at each end.

Arenariomyces: This genus has dark brown carbonaceous perithecia. The asci are deliquescent and the ascospores have three rigid, curved, appendages at each end.

Peritrichospora: The perithecia are elongate and often branched in some species. The perithecial centrum contains large thin-walled polygonal parenchyma connected by cytoplasmic strands. The ascospores are one-celled at first but may become two- to six-celled. They have one rigid, or a fascicle of thread-like appendages at each end, and often a cluster of thread-like appendages at the equator.

Lentescospora: The single species of this genus, *L. submarina*, occurs on decorticated wood in salt or brackish water. The ascospores are one- or two-celled and have small gelatinous caps at each end. The asci are given as bitunicate and deliquescent.

Samarosporella: In this genus the perithecia are immersed in driftwood. The ascospores are one-celled and have a circular wing-like gelatinous envelope.

Trailia: This genus is found on the marine alga *Ascophyllum* and has white fleshy perithecia with an elongate neck. The ascospores are fusoid, rounded above and tapered below into a long filamentous base. They are one- to three-septate and hyaline. Its relationship seems to be with the Nectriaceae.

Haloguignardia: This genus differs from *Guignardia* only in the ascospores which have thickened and extended end walls with a peg-like intrusion of the cytoplasm. The asci are also deliquescent. In *H. decidua* the end walls of the spores are deciduous. Most species occur on the marine alga *Sargassum* and form lobed or branching galls. *Halo-*

guignardia longispora and *H. irritans* form globose to ellipsoid spermagonia containing small ellipsoid spermatia.

Plectolitus: This genus has light colored, non-ostiolate "cleistothecia" seated on a subiculum. The asci are globose to saccate, have a thickened apical wall and are formed at all levels in the ascocarp. Its relationship seems to be with the Plectascales. The ascospores are ovoid, one-celled, hyaline and with numerous spine-like appendages on the surface.

Chadefaudia: The single species *C. marina* occurs on the algal genus *Rhodomynia*. The perithecia are immersed and erumpent as an ostiolar neck. The ascogenous hypha occurs as a spiral multicellular hypha. These cells separate to become the ascogenous cells which separate and lie in a cytoplasmic mass in the center of the perithecium. Each of these cells undergoes nuclear division and gives rise to a thin-walled ascus. The ascus wall is deliquescent and releases the ascospores within the perithecial centrum. This is a developmental characteristic of the Halosphaeriaceae. The ascospores are ellipsoid and have thickened end walls as in *Haloguignardia*, but that genus has a true hymenium.

Paraliomyces: This genus has thick-walled, bitunicate asci. The ascospores are two-celled, hyaline, and have a disc-like, gelatinous appendage at the septum.

Herpotrichiella: The single marine species *H. ciliomaris* has setose perithecia. It differs from the other species of the genus in the twenty to thirty spine-like appendages at each end of the ascospores. The asci are bitunicate.

Ceriosporopsis: The species of *Ceriosporopsis* have erumpent ascocarps and bitunicate deliquescent asci. The ascospores are two-celled and have straight or curved appendages at each end, and in some species several equatorial appendages as well. The ascocarps are supposed to be stromatic in origin. In several species they arise as a parenchymatic stroma. The central interthecial tissues are torn apart into strips as the ascocarp develops, leaving "pseudoparaphyses." The asci are deliquescent.

Antennospora: The perithecia of *Antennospora* are immersed and occur on the roots of *Avicennia marina*. The ascospores have two rigid filiform appendages at each end, which lie at right angles to each other.

Halosphaeria: This genus is held together by the wall structure rather than the ascospores which are two-celled and have a variety of appendage patterns. The outer wall of the perithecium is composed of large, pigmented, rectangular cells; the inner wall of small thin-walled hyaline parenchyma. The asci arise from a basal mass of parenchyma. Hypha-like chains of cells fill the centrum about the hymenium. The ascospores may have a single gelatinous appendage at each end; two to four curved, rigid, hamate appendages at the equator and terminal caps; a hyaline, equatorial, collar-like annulus; an equatorial collar broken up into wedge-like segments and fused with a broad terminal appendage.

Didymosamarospora: This genus has a single species occurring on *Juncus*. The ascospores are two-celled hyaline and surrounded by an irregular undulate sheath.

Lignicola: This genus has dark colored perithecia with ostioles which are short, long or even branched. The ascospores are two-celled without appendages. It is very similar to *Gnomonia.*

Torpedospora: This genus has four- to five-celled ascospores with either four rigid radiating appendages subterminal at each end or terminal at one end.

Orcadia: Orcadia has a white fleshy fruit body with a long ostiolar neck. The asci are fusoid to clavate and curved. The species occur on *Ascophyllum* and *Pelvetia.* The ascospores are fusoid to cylindric, curved, four-celled and hyaline to yellow.

Lulworthia: The species of *Lulworthia* show a wide range of variation in the characters of the ascocarp, which may be light to dark in color and with a short or long ostiolar neck. The perithecial centrum is variable. The ascospores, however, are all filiform, one-celled and hyaline and have conical appendages at each end of the spore. In this respect the species are similar to *Ophiobolus.*

One of the pathogens concerned with the wasting disease of eel grass (*Zostera marina*), which caused the disappearance of this host on both shores of the North Atlantic, was described by Mounce and Diehl (**1934**) as *Ophiobolus marina*. This species has filiform ascospores with apical appendages, however, and has been placed in the genus *Lulworthia.*

Lindra: This genus is also related to *Ophiobolus* in having filiform ascospores, in this case with twenty-five to fifty-two faint septa and globose appendages at each end of the spore. The centrum of the ascocarp is filled with a hyaline parenchyma which disintegrates as the asci are formed.

Maireomyces: This genus includes one doubtful species, *M. personelliae*, which also has long fusoid to filiform one-celled hyaline ascospores, but without appendages.

SELINIACEAE:

Von Arx and Müller (**1955a**) established this family for two genera, *Selinia* and *Seliniella*, which they were unable to place in any established family of the Pyrenomycetes. They believed the two genera to represent primitive types which show relationship to many established orders, but differing in detail in all cases. Both genera occur as saprophytes on dung.

Selinia: This genus has sunken fleshy stromata which are white within and red to brown on the surface and contain one to several perithecia with true walls, an ostiole and paraphyses. The asci are clavate with a simple thickened wall and no apical apparatus. The eight ascospores are ellipsoid, one-celled, hyaline and with a thick gelatinous wall. There are no germ pores or germ slits.

Seliniella: This genus has single immersed, perithecia with the basal wall much thickened. The asci have a simple wall and are surrounded by numerous paraphyses. The ascospores are one-celled, hyaline and with a thickened wall which shows a faintly colored network on the surface at maturity.

HYPOCREALES

The order Hypocreales was originally erected for those species having ostioles and fleshy, bright-colored perithecia or stromata. Many investigators have criticized the use of color as a diagnostic ordinal character. Many genera with light colored perithecia have been placed in or transferred to other orders, and various characters have been used to segregate orders and families from the Hypocreales.

Seaver (**1909, 1910**) recognized two families in the Hypocreales; the Nectriaceae with perithecia borne singly or clustered on an erumpent stroma, and the Hypocreaceae with perithecia immersed within a fleshy stroma. In the Nectriaceae he placed two sub-families, the Nectrieae for species without a stroma and the Creonectriae for those borne on an erumpent stroma. Nannfeldt (**1932**) removed a group of genera which he placed in the Clavicipitales. These genera are all parasitic and have a characteristic long cylindric ascus with a thickened cap with a central canal and long filiform spores. Miller (**1949**) and Luttrell (**1951**) have pointed out that many species in the Hypocreales have a type of "pseudoparaphyses" which grow downward from the apex toward the base of the perithecium, instead of upward, as in the Sphaeriales. Luttrell has used his character as diagnostic for the order Hypocreales. Hanlin (**1961, 1963, 1964, 1965**) has shown this to be true in an additional number of species. There remain, however, many species to be studied as to this behaviour. Von Arx and Müller (**1954**) and Müller and von Arx (**1962**) have recognized two families of hypocreaceous fungi upon the basis of the structure of the ascus tip. They included in the Hypocreaceae all those genera with a simple thickening of the ascus tip. Their Hypomycetaceae have a thickened apical wall of the ascus penetrated by a canal similar to that found in the Clavicipitaceae, but with fusoid spores.

Even though some of the Clavicipitales have a blackened surface to the stroma, others have bright colored perithecia or stromata and they show a relationship to the Hypomycetaceae of Müller and von Arx in the similar type of the ascus tip. The perithecial development of the species of the Clavicipitales, which have been studied, do not show the type of pseudoparaphysis development characteristic of the Hypocreales. They are therefore placed in a separate order next to the Hypocreales.

For a discussion of the taxonomic history, and keys to all of the genera described in the Hypocreales, the reader is referred to the account by Rogerson (**1970**). New Zealand species have been studied by Dingley (**1951-1957a**).

Two families are here recognized in the Hypocreales: Hypocreaceae: Asci with a simple thickened tip, perithecia scattered singly or upon or within a stroma, mostly saprophytic upon dead plant parts or other fungi, some parasitic on higher plants.

Hypomycetaceae: Asci with a thickened apical wall penetrated by a canal, perithecia in or upon a loose byssoid stroma, parasitic upon higher fleshy fungi.

There are many conidial types connected with the ascigerous states of the Hypocreales. They include many form genera of the Moniliales, species which bear their conidia free upon the mycelium or the surface of stromata but not in enclosed locules. The most important of these is the form genus *Fusarium*. This is a large and difficult genus with many species which have been described on many different hosts. They are the cause of important and destructive plant diseases. The conidia are elongate, fusoid, septate, hyaline, tapered at the ends and usually curved or inaequilateral. They have a basal pedicel and are borne singly on simple conidiophores or united into synnemata or compacted masses with agglutinated conidia known as "pionnotes". Microconidia are also formed in many species and thick-walled chlamydospores may occur in the hyphae.

Many species have been described on the basis of host occurrence and many of these have been reduced to synonymy as their ability to infect a wide range of hosts has been shown. The genus is of great importance to and has been studied extensively by plant pathologists. Wollenweber (**1913, 1926**) and Wollenweber and Reinking (**1935**) in studies of the ascus connection have shown that all *Fusarium* species are connected with genera of the Hypocreaceae. Snyder and Hansen (**1940-1945**), in their studies of the genus have reduced the number of species to a dozen or so. The Fusaria are largely soil organisms. They attack the roots and crowns of their host plants and cause wilts and vascular diseases. Infection through the roots results in the proliferation of hyphae in the vascular system and the interference with the transpiration stream resulting in wilting and other pathologic conditions. The production of the lycomarasmin type of toxins also changes the properties of the cellular membranes allowing the leakage of water from the cells. The vascular bundles are often plugged with a red or blackened slime.

Fusarium graminearum causes a scab and foot rot disease of wheat, *F. lini* a wilt of flax, *F. lycopersici* a tomato wilt, *F. vasinfectum* a wilt of cotton and legumes, *F. conglutinans* the cabbage yellows, *F. oxysporum* and *F. cubense* the banana wilt, etc.

Many other hypomycetous form genera, mostly with one-celled hyaline conidia, have also been connected with members of the Hypocreales.

HYPOCREACEAE:

This family, as circumscribed by Müller and von Arx, includes all species which have clavate asci with a simple thickening of the apical wall. It includes the Nectriaceae and part of the Hypocreaceae of Seaver's arrangement. The perithecia may lie free upon or within the substrate without any stroma or upon or within a well developed stroma. Both saprophytic and parasitic species are included and a wide variety of conidial states is known.

The following is a key to the chief genera: (many of these names have been placed in synonymy by some authors)

I. Ascospores 1-celled
- A. Ascospores hyaline
  - 1. Perithecia superficial on host tissue .... PSEUDONECTRIA
    (*Hyponectria* with immersed perithecia is found in the Polystigmataceae .... (Byssonectria)
  - 2. Perithecia superficial upon a stroma.................. ALLANTONECTRIA
- B. Ascospores brown
  - 1. Perithecia upon a stroma .... SPHAERODERMELLA
  - 2. Perithecia immersed within a stroma .... CHROMOCREOPSIS

II. Ascospores 2-celled
- A. Ascospores hyaline
  - 1. Perithecia free; not on or within a stroma
    - a. Perithecia immersed .... NECTRIELLA
      .... (Charonectria)
    - b. Perithecia superficial
      - (I). Ascospores smooth-walled .... NECTRIA
      - (II). Ascospores vertically striate .... NEURONECTRIA
  - 2. Perithecia upon a byssoid subiculum .... NECTRIOPSIS
    .... (Hyphonectria)
  - 3. Perithecia borne on the surface of a stroma
    - a. With a *Stilbum*-like conidial state .... SPHAEROSTILBE
    - b. Without such a *Stilbum* state
      - (I). Perithecia with a much thickened upper wall; on submerged stones ..
        .... HYDRONECTRIA
      - (II). On a fleshy erumpent stroma .... CREONECTRIA
  - 4. Perithecia sunken in a fleshy stroma
    - a. Asci with 16 "half spores"
      - (I). Stroma clavate, stalked .... PODOSTROMA
        .... (Podocrea)
      - (II). Stroma pulvinate to effuse .... HYPOCREA
    - b. Asci with 8 spores
      - (I). With a *Stilbum*-like conidial state .... STILBOCREA
      - (II). Without such a *Stilbum* state
        - (A). Stroma pulvinate to effuse, lobate or radiate ...... HYPOCREOPSIS
        - (B). Stroma globose; perithecia polystichous .... MYCOCITRUS
- B. Ascospores brown
  - 1. Perithecia on the surface of an immersed stroma .... MACBRIDELLA
  - 2. Perithecia sunken in the substrate .... PASSERINULA
  - 3. Perithecia imbedded in a stroma .... CREOPUS
    .... (Chromocrea)
  - 4. Stroma valsoid, immersed .... VALSONECTRIA

III. Ascospores more than 2-celled
- A. Stroma with a *Stilbum*-like conidial state .... STILBONECTRIA
- B. Without such a *Stilbum* state
  - 1. Stroma subglobose perithecia superficial, giving an echinulate appearance
  - 2. Not as above
    - a. Perithecia violet, blue to black .... GIBBERELLA
    - b. Perithecia yellow or red
      - (I). Perithecia on a subiculum .... BYSSOCALLIS
      - (II). Perithecia on a basal stroma .... CALONECTRIA

IV. Ascospores muriform
- A. Perithecia blue to violet .... PLEOGIBBERELLA
- B. Perithecia yellow or red
  - 1. Ascospores hyaline
    - a. Perithecia immersed .... THYRONECTRIA
    - b. Perithecia superficial .... PLEONECTRIA
    - c. With a *Stilbum* state .... MEGALONECTRIA
  - 2. Ascospores brown .... MATTIROLIA
    .... (Thyronectroidea)

Nectria: This genus originally included all those species with superficial perithecia either without or upon a stroma and with two-celled hyaline ascospores. In recent years numerous genera have been segregated upon various characters and the genus limited to free lying perithecia with two-celled, hyaline ascospores. *Nectria sanguinea* (*N. episphaeria*) has bright red superficial perithecia occurring upon the stromata of other fungi. Seaver used the name *Creonectria* for those species having perithecia borne upon a basal erumpent stroma. These species form a stroma within the bark of woody tree species. This stroma ruptures the surface periderm and bears the perithecia upon its surface. These woody species have been studied by Spaulding, Grant and Ayers (**1936**) and by Lohman and Watson (**1953**). *Nectria cinnabarina* (*Crenonectria purpurea*) occurs as a saprophyte on the limbs of many woody trees. It forms erumpent, tan to reddish-brown, pulvinate stromata upon the bark surface on which are borne a hymenium of conidiophores from which one-celled, ellipsoid, hyaline conidia are abstricted. This conidial state is *Tubercularia vulgaris*. The bright red, rough walled perithecia are clustered on this stroma at maturity. The parasitic species upon shade trees has been referred to as *Nectria ditissima*, which is probably a synonym of *N. galligena* (Cayley, **1932**), which is the cause of a canker of apple trees. This pathogen is world wide and causes a canker of the bark. The conidial state of this species is of the form genus *Cylindrocarpon*, which has conidia similar to *Fusarium* but larger, more cylindric, hyaline and one- to six-septate and borne on a sporodochium during the summer. Smaller microconidia which are one-celled or one-septate, are also formed. The perithecia occur on these stromata in late fall.

A bark disease of beech trees which threatened to wipe out this tree in northeastern United States and Canada, was described by Ehrlich (**1934**). It is caused by *Nectria coccinea* var. *faginata* (Lohman and Watson, **1953**). Infection follows damage caused by a scale insect, *Cryptococcus fagi*.

Species with two-celled ascospores, budding in the ascus, have been placed in *Aponectria*; those with polyspored asci in *Metanectria*; those with brown ascospores in *Macbridella*, and those with hairy perithecia in *Lasionectria*.

Nectriella: Müller and von Arx have used this name for species with two-celled, hyaline ascospores and perithecia immersed singly in the host tissue or substrate. Species with an apical ring in the ascus staining blue with KI have been placed in *Leiosphaerella*.

Nectriopsis: This genus has red to violet colored perithecia, with stout ostioles, imbedded in a loose subiculum of light colored hyphae. Conidial states of the form genera *Cephalosporium* and *Fusarium* are known in this genus.

Hydronectria: This monotypic genus was found on submerged stones. It forms a superficial crust of light colored hyphae in cavities containing the perithecia. The perithecial wall consists of a coarse outer layer of parenchyma which is much thickened in the upper portion of the wall, and a thin inner wall of filiform hyphae. It has large, hyaline, one-celled ascospores.

Sphaerostilbe: This genus differs from *Creonectria* only in the conidial state which consists of a stalked synnema on the stroma, of the form genus *Stilbella*. *Sphaerostilbe aurantiicola* is a widespread species found as a parasite on scale insects. The San Jose scale insect has caused great damage to the citrus groves of southern Florida. One method of control is to place branches with scale insects infected with *S. aurantiicola* among the branches of citrus trees affected by the scale insect, thus causing an epidemic of the scale insect. Luttrell (**1944**) has studied the perithecial development of this species. The orange colored perithecia are clustered on a small stroma formed beneath the scale insect. They arise as a short coil at the base of the conidial synnemata in the early spring. This coil enlarges and becomes multinucleate, but no trichogyne, spermatia nor antheridial branch was observed. The coil is surrounded by an envelope of neighboring hyphae, which later pull apart, forming a central cavity. Hyphae from this inner wall grow radially inward. Those on the lower and lateral walls grow slowly and form a thick layer beneath the ascogenous hyphae, which lie at the base of the central cavity. Those from the upper wall grow downeard rapidly forming a palisade-like layer. These are the pseudoparaphyses of the Hypocreales. They press downward against the ascogenous hyphae. The asci arise as deeply staining binucleate cells from these ascogenous hyphae and develop from typical croziers. They push upwards between the pseudoparaphyses which serve as a nurse tissue.

Hanlin (**1961**) has reported on the development of *Nectria gliocladioides*. The perithecium in this species arises as a swollen uninucleate tip of a vegetative hypha. This cell becomes multinucleate, septate and branches to form a coiled archicarp. No fusion of hyphae or cells was seen. A proliferation of hyphae from its basal cell forms an enveloping wall. The enlarged ascogonial cells of the archicarp divide to form smaller cells. The outer wall is continuous with the basal stroma. The inner basal wall divides to form a mass of "subhymenial parenchyma." A meristematic region originates at the inner apex of the perithecial cavity. This meristem produces a mass of small cells which elongate downward. These cells continue to elongate to form a palisade of pseudoparaphyses which push the ascogenous cells toward the subhymenial parenchyma. The asci arise from this crescent shaped layer of ascogenous cells and grow upward through the pseudoparaphyses. The asci were thin-walled with a thickened lateral wall and an apical ring which is not amyloid, with a funnel-shaped perforation in the center. This species has a *Verticillium* state consisting of verticillately branched conidiophores and a *Gliocladium* state with apical, penicillate, phialides on the conidiophores.

The genus *Nectria* has a wide variety of conidial states connected with its ascigerous states. They include the form genera *Fusarium, Heliscus, Microcera, Acrostalagmus, Cylindrocarpon, Gliocladium, Verticillium, Cephalosporium, Tubercularia, Stilbella, Dendrodochium, Spicaria, Graphium*, etc. In *N. ralfsii*, Dingley (**1957**) found a conidial state of the genus *Dendrodochium*, consisting of stromatic sporodochia with penicillately branching conidiophores bearing colored, one-celled, fusoid conidia. In *N. otagensis* an *Aschersonia* state having an open labyrinthiform cavity in a stroma which produced catenulate conidia was seen. In *N. macrostoma* a *Graphium* type of dark colored, stalked synnema with a head of catenulate hyaline conidia was formed.

Neuronectria: Munk (**1957**) erected this genus for *Nectria peziza*, which has superficial perithecia. It differs from most *Nectria* species in the thin-walled cells of the perithecial wall, the lack of paraphyses and the longitudinal striae on the ascospores. These conclusions were supported by Hanlin (**1963**) who studied the development of the perithecia. The ascocarp originates as the swollen septate tip of a hypha, which was surrounded by a deep staining hypha. This swollen tip continued to divide to form a coiled multinucleate archicarp. No fusion of hyphae or spermatia was observed. This archicarp was enveloped by surrounging hyphae to form a perithecial wall. At this stage the centrum of the primordium was filled with large deeply staining cells of the archicarp. The inner wall cells at the ostiolar end of the perithecium became meristematic and deeply staining. The central archicarp cells became thin-walled and vacuolate, except for a few deeply staining ones at the bottom of the cavity. The meristematic cells at the ostiolar end began to divide and form a mass of isodiametric pseudoparenchyma. These cells became arranged in vertical rows and pushed downward the ascogenous cells toward the base of the perithecium. At this stage the upper portion of the perithecial cavity was filled with this isodiametric pseudoparenchyma (the "supra-hymenial parenchyma"). The lower third of the cavity was filled with the large disintegrating ascogenous cells ("subhymenial parenchyma") and between these two layers there lay the crescent shaped layer of ascogenous cells from which the asci arose. As the asci grew upward some of the central cells elongated into filaments and others disintegrated. Hanlin did not look upon these rows of pseudoparenchyma cells as true pseudoparaphyses. The ascospores furthermore are two-celled and hyaline but show definite vertical striae.

He accepted Munk's genus *Neuronectria* upon the basis of these characters. He suggested that the development of the perithecium might place the genus in the Melanosporaceae.

Although the rows of sterile tissues consist of rows of parenchyma cells rather than elongate filaments, the origin of this tissue is identical to that of true pseudoparaphyses.

Valsonectria: Valsonectria has a light colored valsoid stroma immersed in woody stems. The perithecia are clustered within the stroma and have elongate ostioles which are collectively erumpent. The ascospores are two-celled, light brown and have vertical striae as in *Neuronectria.* In the colored valsoid stroma it resembles *Endothia*, but the ascospores are distinct Loeffler and Müller (**1961**) report a *Sporotrichum* state for *V. reticulata.*

Hypocreopsis: This genus has a pulvinate to effuse, often lobed, or radiate stroma with the perithecia immersed within it, as in *Hypocrea.* The ascospores, however, are ellipsoid to fusoid and do not separate into part spores. They are hyaline to light yellow. The stromata may be light colored or with a brown or black crust. The species with a dark colored surface have been placed in *Myrmaeciella* or *Porphyrosoma*. Some species of *Hypocreopsis* have a *Stilbum*-like conidial state consisting of stalked heads of conidia (*Stromatocrea*). Such species would fall in the genus *Stilbocrea* which is based upon such a conidial state.

Mycocitrus: Mycocitrus has large bright colored stromata somewhat as in *Hypocrea*, but with the perithecia imbedded in a polystichous manner, in several layers. The ascospores are two-celled, hyaline, but do not separate into part spores.

Hypocrea: Hypocrea has pulvinate to flattened, effuse, fleshy, bright colored stromata with the perithecia imbedded in the upper layers with erumpent ostioles. The species are saprophytic upon decaying wood. The ascospores are ellipsoid, two-celled, hyaline and the two cells separate in the ascus to form sixteen globose to cuboid "partspores." Some species have olive-green ascospores and these have been placed in the genera *Creopus* or *Chromocrea*. The conidial states of *Hypocrea* belong to the form genus *Trichoderma*, which forms greenish colored masses of globose conidia on branching conidiophores.

Hanlin (**1965**) has reported upon the perithecial development of *Hypocrea schweinitzii* (*H. lenta*). This species has olive green pulvinate stromata. The ascocarps arise as a coiled, septate archicarp of uninucleate cells, just beneath the colored surface layer of the stroma. These cells become multinucleate, but no cellular fusion occurs. A proliferation of hyphae from the basal cells of this coil envelope this primordium. The hyphae on the inner surface of this envelope become meristematic and grow radially inwards. The filaments at the base of the cavity soon stop their growth in length and divide to form a layer of parenchyma. The filaments form the upper surface of the wall layer, continue to elongate downwards to form a compact palisade of hyphae, the pseudoparaphyses. Meanwhile, the ascogenous cells of the archicarp divide to form a thin layer of small darkly staining ascogenous cells which are pushed downward by the developing pseudoparaphyses. As the asci develop from these ascogenous cells and push upward through the pseudoparaphyses, these filaments disintegrate, acting as a nurse tissue. This is the typical development of the perithecium in the Hypocreales.

Podostroma (*Podocrea*): This genus has sixteen globose part spores derived from the eight two-celled ascospores, as in *Hypocrea*. The stroma, however, is clavate with a basal sterile stalk, is light colored and occurs on soil.

Macbridella: This genus has perithecia on a basal stroma as in *Creonectria*, but two-celled ascospores which are olive green in color.

Passerinula: This genus has two-celled dark colored ascospores and perithecia which are imbedded in the substrate.

Creopus (*Chromocrea*): This genus has perithecia imbedded in a stroma and ascospores which break up into sixteen part spores, as in *Hypocrea*, but the ascospores show various degrees of pigmentation. Dingley (**1952, 1957a**) includes the genus in *Hypocrea*.

Creopus spinulosa was studied by Doguet (**1957**). The ascocarp arises as a branching multinucleate hypha, the ultimate branches of which form a coil. The surrounding stroma forms a double wall, an outer layer of parenchyma which gives rise to the ostiole and an inner layer of plectenchyma including the archicarp. This inner layer of plectenchyma differentiates into a lower layer of inflated parenchyma, the "sus" or subhymenial parenchyma of Douget and an upper portion, the cells of which elongate and grow downward to form the pseudoparaphyses, which in turn push the ascogenous cells, derived from the archicarp, against the "sus parenchyma". This development conforms with that described by Hanlin for *Hypocrea*.

Gibberella: The genus *Gibberella* has two- to many-septate ascospores which are hyaline or somewhat colored. The perithecia are colored blue to violet or dark, and may be borne singly or in a subiculum or on a basal stroma.

Gibberella zeae is the cause of a scab, seedling blight, or root rot of wheat. The fungus grows on stalks and straw of the grain as a saprophyte, and infects the flowering heads where it forms gray to pinkish mats of mycelium. The mycelium grows into the developing ovules and causes a shrinking and wrinkling of the gray to brown grain, known as "scab". The starch of the grain is reduced to sugars and fatty acids. This scabby grain may be toxic to animals, especially pigs, and less so to cattle and chickens. The conidial state, *Fusarium graminearum*, occurs on the surface mycelium of the grain. The perithecia occur later in the year on the grain and on the stubble in the fields. Seedling blight and root rot result from the infection by ascospores from the stubble in the field.

Gibberella fujikuroi causes a disease of rice. It is the source of gibberellic acid which causes an elongation of the nodes of the rice plant. The conidial state is *Fusarium moniliforme.*

Calonectria (*Scoleconectria*): In this genus the ascospores have more than one transverse septum. They may be ellipsoid, fusoid or elongate cylindric and hyaline or somewhat colored. The perithecia are yellow or red in color and are borne on a basal stroma. Species with elongate ascospores approach those of the Clavicipitaceae, but do not have the cylindric asci and apical thickened cap of that family, and are saprophytic.

Thyronectria: This genus includes all those species of the Hypocreaceae with muriform ascospores and perithecia not violet or blue. The type of stromatic development and spore color are variable and a number of genera have been segregated as indicated in the generic key, but Seeler (**1940a**) monographed the genus and includes them all in *Thyronectria.*

Seeler (**1940**) found *T. austroamericana* to be the cause of a wilt disease which killed trees of *Gleditsia japonica* in the Arnold Arboretum and of *G. triacanthos* on Nantucket Island. The fungus mycelium penetrated the wood and caused a discolored precipitation of gummy material in the first annual ring of the wood vessels. This blocks the transpiration stream and causes a withering of the leaves and eventual defoliation. The infected trees have an exceptionally large crop of fruit before their death. The fungus is endemic to America, and the Asiatic *Gleditisia* is more susceptible than the American *G. triacanthos*. Dead wood of *G. japonica* produced fruiting stromata in a moist situation. Pycnidia containing minute cylindric, hyaline conidia were formed. Perithecia occurred later in the fall. The perithecia arose as a group of large multinucleate cells, the archicarp, which produces a filamentous hypha, the trichogyne, which penetrated to the surface of the stroma. Seeler saw conidia attached to the trichogyne, but no nuclear fusion was observed. A perithecial wall was formed around the archicarp. Hyphae grew downwards from the roof of this wall, pressing the ascogenous cells towards the base. This is the typical development of pseudoparaphyses as found in other members of the Hypocreales.

HYPOMYCETACEAE:

The family Hypomycetaceae was originally differentiated from the Nectriaceae and Hypocreaceae on the presence of a basal subiculum of loosely interwoven hyphae, rather than a fleshy stroma. Munk (**1957**) emphasized the character of the ascus tip which has a thickened apical wall through which a canal runs to the exterior. This is somewhat similar to that of the ascus tip found in the Clavicipitaceae, which differs in the filiform ascospores. The members of this family are found mostly as parasites upon the higher fleshy fungi.

The few genera may be keyed out as follows:

I. Ascospores 1-celled.. PECKIELLA
II. Ascospores 2-celled
  A. Ascospores fragmenting into the two cells .... ARACHNOCREA
  B. Ascospores not so separating
    1. Perithecia small with an elongate ostiole; asci clavate .... PYXIDIOPHORA
    2. Perithecia larger, ostiole short; asci cylindric .... HYPOMYCES

Peckiella: This is the only genus in this family with one-celled spores.

Hypomyces: Hypomyces is the largest genus. The species occur as parasites on the basidiocarps of the Boletaceae and Agaricaceae, usually fruiting upon the decaying fruit bodies of these fungi. The perithecia lie in a superficial byssoid subiculum. The asci are cylindric and have an apical cylindric cap-like wall thickening, penetrated by a canal. The ascospores are two-celled, fusoid, hyaline or light colored and with a short tapered appendage (apiculus) at each end. Species with ascospores which are unequally two-celled have been placed in a genus *Apiocrea*. The subiculum of some species becomes colored yellow from the presence of numerous echinulate chlamydospores of the form genera *Sepedonium* (one-celled) or *Blastotrichum* (septate). Hypomycetous conidial states with one- to several-celled conidia of the form genera *Trichothecium, Dactylium, Verticillium,* and *Diplocladium* have also been reported for this genus.

Hypomyces lactifluorum is an interesting species which is parasitic upon the fruit bodies of species of *Russula* and *Lactarius*. It apparently infects the mycelium, for very young fruit bodies are already misshapen, forming a contorted fruit body with a stromatic orange-colored layer replacing the gill surface. Numerous perithecia are imbedded in this stroma. Hanlin (**1962**) has followed perithecial development in this species. The perithecial initial consists of a spirally coiled, septate hypha with as many as ten spirals and uninucleate cells. These cells become multinucleate and branched to form a central tissue around which enveloping hyphae form an outer wall. Certain of the archicarp cells divide to form a layer of smaller cells in the base of the perithecial cavity. A meristem develops at the apex of the perithecium and these cells grow downward and press the ascogenous cells toward the base of the cavity. These are the pseudoparaphyses as found in *Nectria, Thyronectria, Hypocrea,* etc. Two-celled and three-celled echinulate chlamydospores of the form genus *Blastotrichum* were found on the subiculum hyphae.

Hanlin (**1964**) also studied *Hypomyces trichothecoides*. The perithecial initials here were also tightly coiled hyphae with up to ten turns. They were divided into uninucleate cells. Further development was like that in *H. lactifluorum*, with the large cells of the archicarp dividing to form a layer of smaller cells. The apical meristem, here, also grew downward as a palisade of filiform pseudoparaphyses. A conidial state of the form genus *Trichothecium* with one- or two-celled, hyaline, cylindric conidia, was formed.

Hypomyces solani forma *cucurbitae* (formerly *Nectria* (*Hypomyces*) *ipomoeae*, Cook, **1923**), with the conidial state *Fusarium javanicum*, causes the root and fruit rot of cucurbits and other plants. It has been studied as regards its genetics by several authors. Dimock (**1937**) found this species to be heterothallic. All strains produced small reddish perithecial primordia, but these did not mature unless fertilized by spermatia from a compatible strain (A or a). Normal strains produce macroconidia in blue, green or brown sporodochia, and also micro-conidia. Snyder (**1940**) reported a strain which produced a white strain (Wa) were placed on a red strain (RA) all the perithecia were red but the ascospores from the red hybrid perithecia gave a one:one ratio of red to white strains, indicating that the mutation was due to a genetic factor, and that the perithecial wall was vegetative tissue of the female (perithecial) parent.

Hansen (**1938**) had previously shown that in many Fungi Imperfecti there is a general tendency to produce mutants called the "dual phenomenon." These were called the C or conidial and the M or mycelial types. The C strains produce conidia, but the M strains do not, remaining mycelial only. These types were genetically stable. M types arose from C types by mutation, but the reverse mutation, C from M, did not occur. In 1943 Hansen and Snyder found this condition to occur in *Hypomyces solani* forma *cucurbitae*. The C and M genes were independent of the sex genes A and a. Hirsch (**1949**) found that there were four types of mutant strains in this species, namely hermaphrodite, female, male, and neuter. After some discussion as to the genetic situation in the species, El-Ani (**1954, 1956**) proposed the following explanation:

There are four double chromosomes in the zygote and four haploid chromosomes in the mycelial strains. There are three pairs of allelomorphs as follows: the sex genes A and a in one chromosome, the genes C and c which govern the production of perithecial initials and macroconidia, the genes M and m which govern the formation of spermatia but no conidia or perithecial primordia, in other chromosomes. The properties of the genes are as follows: (A and a are both necessary for the formation of a zygote).

The gene C allows the formation of perithecial initials and macroconidia.
The gene c does not give perithecial initials nor macroconidia.
The gene M allows the formation of mycelia and microconidia but not perithecial initials nor macroconidia.
The gene m gives only mycelia and no other structures.
The genes C and c, and M and m are on the same pair of chromosomes but not on the same chromosome as the sex genes A and a.

The following strain combinations may occur:

CMa and CMA: hermaphrodites with perithecial initials, macroconida and spermatia. Any strain can fuse with any compatible strain.

Cma and CmA: these are female, producing perithecial initials and macroconidia but no spermatia; they can cross only with compatible hermaphrodite or male strains.

cMa or cMA: these are male strains, producing spermatia but no perithecial initials nor macroconidia; they cross only with hermaphrodite or female strains.

cma or cmA: these are neuter strains, producing only mycelium; no perithecial initials, macroconidia, or spermatia are formed; they are sterile in all crosses. Both perithecial initials and spermatia with compatible nuclei are necessary for fertile crosses.

A mutation in either a male or a female strain gives rise to a neuter. The mutation of a male or female gene in a hermaphrodite gives rise to a male or female strain (mutations are always the loss of a character).

Pyxidiophora: The perithecia of *Pyxidiophora* are small and have an elongate penicillate ostiole. They are parasitic upon the fruit bodies of the parasitic agaric *Nyctalis*, lying in a surface subiculum or sunken in the pileus. The asci are clavate but with a thickened apical plug penetrated by a canal, and contain two to six two-celled, hyaline ascospores with a short basal appendage.

Arachnocrea: This genus has light colored perithecia borne on a basal subiculum on agaric fungi. It differs from *Hypomyces* only in the fact that the two-celled fusoid ascospores fragment into the two component cells.

CLAVICIPITALES

There has been a good deal of controversy as to the relationship of the Clavicipitales. The genera placed here were originally placed in the Hypocreales because of the fleshy, bright colored, stromata. Certain species, as in *Balansia* however, have a blackened surface crust and many authorities have criticized the use of color as an ordinal character.

The members of this order are all parasites, either upon grasses and sedges or on insects and other fungi. The perithecia may be superficial upon a subiculum or stroma or imbedded within a stroma, which may be bright colored or blackened. The asci are mostly long cylindric and have a much thickened ring-like cap penetrated by a canal. This cap is deciduous when the ascospores are ejected. The ascospores are long cylindric, one-celled or septate, and fragment into their component cells in many species. In the species studied, there are no pseudoparaphyses formed as in the members of the Hypocreales. The perithecial development is rather distinct, as indicated in the following examples.

Jaczewski (**1894**) claimed that these genera did not have a true perithecial wall, and placed them in the Dothideales. Vincennes (**1917**) claimed there was no perithecial wall in *Epichloe* and *Claviceps* and that the asci were borne in mere cavities in the stroma. On the basis of these conclusions, Nannfeldt (**1932**) erected the Clavicipitales as an order, transitional between his Ascoloculares and Ascohymeniales. On the other hand, Jenkins (**1934**) described a true wall in *Cordyceps agariciformia*, as did Doguet (**1960**) for *Epichloe*. Miller, Luttrell and Munk also agreed that there was true wall in these genera and placed them as a family in the Sphaeriales. Luttrell placed this family near the Xylariales and Munk placed it next to the Hypomycetaceae.

In the opinion of the present writer, the Clavicipitales are related to the Hypomyceteae because of their parasitic habit and the thickened apical cap with its central canal and the predominantly light colored stromata. They differ in the long filiform ascospores and the development of the perithecium, which does not have the pseudoparaphyses used by Luttrell as a diagnostic ordinal character of the Hypocreales. These genera are, therefore, placed as an order next to the Hypocreales. The Xylariales differ in the one-celled brown ascospores with germ pores or germ slits and asci with an amyloid ring in the apex. The Diaporthales differ in their lack of a fleshy stroma, dark colored perithecial walls and lack of filiform spores and asci with a refractive ring in the apex.

This is a striking example of the number of hypothetical relationships possible as a result of the emphasis of different single characteristics. (Any arbitrary taxon will have marginal members with characters suggesting relationships with neighboring taxa.)

The order is divided into two families on the basis of their type of host. The Clavicipitaceae are parasitic upon grasses and sedges, whereas the Çordycipitaceae are parasitic upon insects and other fungi.

CLAVICIPITACEAE:

The members of this family are parasitic upon grasses and sedges. They form a sclerotial basal stroma within some portion of the host which bears the conidial state and upon or from which the perithecial stroma arises. Diehl (**1950**) has separated the genera upon the type of conidial state which they possess. A key to the genera follows:

I. Perithecial stromata arising from a free lying deciduous sclerotium; conidial state *Sphacelia* .... CLAVICEPS
II. Perithecial stromata upon or from an effuse básal sclerotium in leaves, stems or florets
  A. Sclerotium and perithecial stroma effuse
    1. Perithecial stroma bright colored; conidial state a *Typhodium* ..... EPICHLOE
    2. Perithecial stroma with a blackened crust; conidial state an *Ephelis* .... DOTHICHLOE
  B. Sclerotium effuse; perithecial stroma pulvinate or stalked
    1. Conidial state an *Ephelis* .... BALANSIA
    2. Conidial states of two types: *Ephelis* and *Typhodium* ...... ATKINSONELLA
    3. Conidial state ustilaginoid............................ USTILAGINOIDEA
    4. No conidial state present .... BALANSIOPSIS

Claviceps: Claviceps parasitizes the inflorescences of many grasses. The best known species is *Claviceps purpurea*, the ergot of grains. The infecting mycelium enters the ovary of the grass flower and turns it into a sclerotium. At first the surface of the sclerotium is irregular chambered and covered with a hymenium of conidiophores bearing ellipsoid, one-celled, hyaline conidia which are exuded in a sweet sticky nectar, known as "honey dew," which is fed upon by insects which serve to distribute the conidia. This is the *Sphacelia* state which is characteristic of the genus. As the season advances, the sclerotium becomes compacted into a pyriform, purplish-black structure which falls out of the inflorescence. This sclerotium is the "ergot" of commerce and is the source of a number of drugs like histamine, ergotamine, egosterol, ergotoxin, etc. The largest type of ergot is produced on rye and is obtained by artificial inoculation as a commercial source. The chemicals in the ergot are toxic to man and animals and if introduced into the flour made from grain as a contaminant can cause a severe illness. Such contaminations have caused plague-like epidemics throughout central Europe where rye flour is used. These sclerotia fall to the ground and after a period of rest at a low temperature can germinate to form a group of small stalked stromata with a globose, pinkish head, in the surface of which the perithecia are imbedded. The asci are cylindric with the characteristic thickened apical cap, and the ascospores are long filiform, non-septate in the ascus, but becoming septate upon germination.

The development of the perithecia in *Claviceps purpurea* has been followed by Killian (**1919**). When the sclerotia germinate, the stipe breaks through the cortex and expands at its apex to form the globose head. The perithecia originate in a hyphal layer just beneath the surface crust of the head as vertically elongate, darkly stained, hyphae, the archicarps. These archicarps occur in groups. The apical portion of the hypha becomes enlarged and multinucleate. Such hyphal tips form two or more branches, the outer of which grow up around the swollen central tip and eventually fuse with it. Nuclei migrate from this "antheridial" branch into the "ascogonial" cell. The lower stalk portion of the archicarp becomes septate and the swollen upper portion with its nuclei degenerates. The lower cells, which have paired nuclei, proliferate to form the ascogenous hyphae. This archicarp is similar to that reported for *Epichloe*, by Doguet (**1960**), except that there was no fusion of cells reported for that species. As the ascogenous cells proliferate, the neighboring hyphae of the stroma orient themselves around them to form the perithecial wall. The inner hyphae of this wall grow inward radially into the cavity to form the true paraphyses of the perithecium. The upper portion of these hyphae grow up through the crust of the stroma to form the ostiole. The asci arise from binucleate branches of the ascogenous hyphae or as croziers and dissolve their way between the paraphyses as they mature. This is a type of development which seems to be common to the Clavicipitales.

Epichloe: The genus *Epichloe* is parasitic on grasses and forms effuse colored stromata on the stems and causes an elongation of the nodes. The common species, *E. typhina*, is systemic in the plant and the stroma originates above the nodes while they are still beneath the leaf sheath. The stalk above the node remains sterile. The stroma at first forms a thin white layer covered by a layer of clavate, phialid-like conidiophores bearing ellipsoid, one-celled, hyaline, dry conidia, known as the *Typhodium* state.

Doguet (**1960**) has studied perithecial development in this species. The perithecial primordium arises in the surface layer of the young stroma as a vertical deeply staining hypha which becomes spirally coiled at the apex. This apical coil increases in size and the basal cells disintegrate. Certain hyphae are differentiated from its base and grow upward about the archicarp coil. These hyphae form a perithecial wall about a central cavity. Lateral branches from these hyphae grow inward into this cavity. These hyphae give rise to the paraphyses within the perithecial cavity, and at the apex to the periphyses of the ostiole. A layer of stromatic tissue hyphae differentiates to form an outer wall of dark, thick-walled cells. Meanwhile the mass of ascogenous cells proliferates and gives rise to the ascus primordia which grow up between the paraphyses and dissolve them away. It can be seen that there is no downward growth of pseudoparaphyses in this species, but rather an inward and upward growth of true paraphyses. Doguet paid special attention to the apex of the ascus in this species. He claimed there is a double wall. The outer wall is thin and modified at the apex as a "calotte," as described by Chadefaud. The inner wall becomes thicker as the ascus develops. The thickened cap he described as a ring, or "anneaux," appearing as two thickened caps in section with a canal between, similar to that in the Diaporthaceae. At maturity the inner wall swells and the upper portion with the cap breaks off and the spores are pushed out. He suggested that the ascus might be considered bitunicate.

Balansia - Dothichloe: Diehl (**1950**) has monographed *Balansia* and related genera. These genera have an effuse basal sclerotium or hypothallus, upon which the perithecial stromata are formed. Diehl separated the genera upon the type of conidial state produced. He included all species with an *Ephelis* conidial state only in the genus *Balansia.* This includes *Dothichloe* and *Ophiodothis* as synonyms. These species occur as parasitic on the stems, leaves and florets of grasses and sedges. In *Dothichloe* the *Ephelis* state consists of hymenia in hysteriform cavities in the young hypothallus containing long filiform hyaline conidia. The perithecial stromata are sessile, effuse or pulvinate upon the hypothallus and with a blackened surface crust. In *Balansia* the *Ephelis* state forms hymenia in flattened, patellate or cup-like cavities with similar filiform conidia. The perithecial stromata are globose, sessile or stipitate upon the hypothallus and may be black or yellow in color. Diehl found that there are intermediates in these characters and placed them all in the same genus. Infection may be systemic and the mycelium proliferates in the ovules and is distributed in this manner. Infection usually causes sterility in the host plant. The asci are cylindric with a thickened apex penetrated by a canal. The ascospores are long filiform, hyaline and break up into segments within the ascus.

Aktinsonella: This is a genus erected for those species in which both *Ephelis* and *Typhodium* conidial states occur. In all other respects it is similar to *Balansia.*

Balansiopsis: This genus includes those species in which no conidial state is known. Otherwise it is also similar to *Balansia.*

CORDYCIPITACEAE:

This family as used here is similar to the tribe of the same name mentioned by Seaver (**1911**) and Diehl (**1950**) but includes all those species with cylindric asci and filiform spores which are parasitic upon insects., spiders and other fungi. The mycelium penetrates the host and forms a sclerotium within the host from which there arises a subiculum or stroma bearing first conidia and later the perithecia. The conidia are mostly borne on synnemata. The perithecia may occur on a subiculum or upon or in a pulvinate, massive or stipitate stromata often in a clavate or globose head.

Cordyceps: This is a most interesting genus, nearly all species of which occur on insects or spiders, except for a few species on the subterranean fungus *Elaphomyces*. The asci are long cylindric with a thickened apical cap-like wall penetrated by a canal. The ascospores are long cylindric, hyaline, many-septate and fragment into their component cells. Species occur on all stages of insects, lavae, pupae and mature adults. The mycelia penetrate the body of the insect and turn it into a sclerotical mass. From this sclerotium there arises a conidial state which usually consists of a stalked synnema with conidiophore hyphae produced in a head or along the sides of the synnema, bearing various conidia. This stalk later expands to form a clavate or globose head within or upon which the perithecia develop. Conidial states in the form genera *Isaria, Hirsutella. Hymenostilbe, Akanthomyces, Stilbum. Insecticola, Tilachlidium*, and *Synnematium*, have been reported. These genera have been discussed by Mains (**1950, 1951**). Some species have hyphomycetous conidial states as in *Spicaria, Sporotrichum, Cephalosporium,* etc. Species of *Cordyceps* have been described by Seaver (**1911**), Massee (**1895**) and Y. Kobayasi (**1941**). Mains (**1934-1959**) has published descriptions and excellent photographs of many species.

In *Cordyceps agariciformia*, W. Jenkins (**1934**) found that the perithecia arise in the interzonal area just beneath the cortex of the fertile head as a three- to five-septate, slightly coiled hypha. These archicarps arise in groups and become more coiled and multinucleate. No fusions were seen. Neighboring hyphae of the stroma envelope and grow between these archicarp hyphae, which eventually come to lie in a cavity. The inner hyphae of this wall become loosened and grow inward to form the paraphyses within the cavity. Some of the cells of the archicarp form the ascogenous hyphae in the base of this cavity. The other cells of the archicarp become empty plectenchyma on the cavity floor. The asci arise as simple branches or croziers of the ascogenous system.

Torrubiella: This genus is found chiefly on scale insects and is similar to *Cordyceps*, but the perithecia are borne superficially upon a byssoid subiculum, and not on a stalked stroma. The asci are cylindric with the characteristic apical plug and the ascospores are filiform, septate, and fragment into part spores.

Ophionectria: This genus was described by Lloyd for a fungus similar to *Torrubiella* but parasitic upon the stromata of *Cordyceps*. The perithecia are superficial upon flake-like stromata on the fertile heads of *Cordyceps*.

Ophiocordyceps: This is a genus proposed for species with stromata similar to those of *Cordyceps* but with asci having a fusoid apex with a poorly developed thickened apical cap. The ascospores are elongate fusoid, overlapping in the ascus, and do not fragment into part spores.

Podonectria: This genus, on scale insects, has superficial perithecia upon a byssoid stroma, as in *Torrubiella*, but the asci are cylindric to fusoid with a poorly developed apical cap and the ascospores are long fusoid at one end, multiseptate and biseriate in the ascus.

Barya: The species of this genus have superficial perithecia upon a byssoid stroma. The ascospores are non-septate, in contrast to *Torrubiella*. The species are parasitic upon other fungi (or lichens).

Oomyces: The species of this genus are also superficial on a basal stroma, but each perithecium is enclosed in an eveloping stroma, forming a double wall.

Hypocrella: This genus forms massive stromata enveloping the host, on white flies (*Hypocrella*) or scale insects (*Fleischeria*). Within this stroma a large pycnidial state of the form genus *Aschersonia* is formed, containing cylindric to fusoid conidia. The perithecia are imbedded in the surface of this stroma. The asci are cylindric with a thickened apical cap and the ascospores are long cylindric, multiseptate and fragmenting into part spores in the ascus. The species of *Aschersonia* are described by Mains (**1959a**).

LABOULBENIALES

The Laboulbeniales are an interesting and highly specialized group of parasites upon the exoskeleton of insects and a few arachnids. Some 1500 species are known and most of them are limited to certain genera or species of insects. Some species are even limited in their occurrence to certain portions of the insects anatomy. Benjamin and Shanor (**1952**) in their study of six species of *Laboulbenia* on beetles, have shown that some species occur on both male and female individuals whereas others are limited not only to male or female insects but also occur on very limited areas of the insects body. Thaxter (**1896**), in his series of exhaustive monographs of this order, suggested this same phenomenon. This distribution seems to be, in part at least, dependant upon the position of the insects during copulation, when the transfer of ascospores and infection take place.

The ascocarps of these fungi are very small, 0.5-3 mm tall, and dark in color. They show a wide variety of modifications of the basic pattern. They normally consist of a cellular "receptacle," which bears sterile appendages, antheridial branches and perithecial stalks. Some species are monoecious with both antheridia and perithecia upon the same plant and others are dioecious with the perithecia and antheridia on separate individuals.

The development of the fungus follows a basic pattern in all of the species studied. This development has been followed in great detail, describing all of the individual cell divisions and progression in a number of genera, such as *Stigmatomyces* (Thaxter, **1896**), *Laboulbenia* (Thaxter, **1896**; Faull, **1912**; and Benjamin and Shanor, **1950, 1950a**), *Filariomyces* (Shanor, **1952**) and *Euzodiomyces* (Benjamin and Shanor, **1951**).

The ascospores of the Laboulbeniales are long fusoid, two-celled and hyaline, with the upper cell longer than the lower one, and surrounded by a gelatinous envelope, which persists as a sheath about the developing thallus. In dioecious species the ascospores are ejected in pairs and one spore of the pair develops into a male and the other into a female plant. Upon germination, the lower end of the longer cell becomes modified into a blackened "foot" which penetrates into the exoskeleton and functions as an absorbing organ. Further division of this longer cell, which is now the basal one of the thallus, gives rise to the receptacle, whereas septation of the upper smaller cell forms the primary and secondary appendages of the thallus. The antheridia usually arise as branches of the primary appendage, and the perithecia from a cell of the receptacle. Spermatia are formed exogenously as buds from the tips of secondary appendages in the family Ceratomycetaceae and endogenously from phialid-like antheridia in the other two families of the order. In the Laboulbeniaceae, the spermatia are ejected from antheridia borne separately on the branches. In the Peyritschiellaceae the antheridia are in a compound structure and empty into a common cavity. These spermatia were supposed, by Thaxter and others, to fuse with the trichogynes of the perithecium and bring about fertilization. The "procarp" or primordium of the perithecium arises by the division of a cell of the receptacle. The "archicarp" or female sexual apparatus consists of a three-celled structure, consisting of a basal ascogenic cell, a central trichophoric cell, and a terminal trichogyne which may become very elongate or even branched. The nuclei of the ascogenic cell and the trichophoric cell undergo a series of divisions resulting in a four nucleate cell which divides longitudinally into two "ascogenous cells" which are binucleate and bud out the young asci.

Thaxter (**1896-1931**) has published a five volume monograph of this order with magnificent illustrations and descriptions of the great majority of the known species. He recognized three families as follows:

I. Antheridia lacking, spermatia budding off exogenously..... CERATOMYCETACEAE
II. Spermatia formed endogenously from phialid-like antheridia
  A. Antheridia borne singly on branches of the appendages .... LABOULBENIACEAE
  B. Antheridia compound; several opening into a common chamber............. .... PEYRITSCHIELLACEAE

LITERATURE CITED

Allen, R. F. 1936. A cytological study of *Erysiphe polygoni* on *Delphinium.* J. Agr. Res. **58**: 801-818.

Ames, L. M. 1961. A monograph of the Chaetomiaceae. U.S. Army Res. Dev. Ser. No. **2**: 1-125.

Andrus, C. F. 1936. Cell relations in the perithecium of *Ceratostomella multiannulata.* Mycologia **28**: 133-153.

—— **& L. L. Harter.** 1933. Morphology and reproduction in *Ceratostomella fimbriata.* J. Agr. Res. **46**: 1059-1078.

Arnaud, G. 1910. Contribution a l'étude des Fumagines. Ann. Ecole nat. Agric. Montpelier, Sér. 2, **9**: 239-277.

——. 1925. Les Astèrinées, IV. (Étude sur les Champignons Pyrenomycetes). Ann. sci. nat. Bot. Sér. 10, **7**: 643-723.

——. 1931. Les Astérinées, VII. Mus. nat. Hist. Nat. Ann. Crypt. Exot. **4**(2): 74-97, pl. 2-7.

Arnold, C. A. 1928. The development of the perithecium and spermagonium of *Sporormia leporina* Niessl. Amer. J. Bot. **15**: 241-245.

Arnold, R. H. & R. C. Russell. 1960. *Cucurbitaria staphula* and its association with *Macrophoma tumefaciens.* Mycologia **52**: 499-512.

Arx, von J. A. 1949. Beiträge zur Kenntnis der Gattung *Mycosphaerella.* Sydowia **3**: 27-100.

——. 1951. Über die Gattung *Laestadia* und die Gnomoniaceen. Anton V. Leeuweenhoek **17**: 259-272.

——. 1952. Studies on *Venturia* and related genera. Tijdschr. Plantenziekt. **58**: 260-266.

——. 1962. Über die Ascomyceten-Familien der Leptopeltaceae. Acta Bot. Neerland **13**: 182-188.

——. 1963. Die Gattungen der Myriangiales. Persoonia **2**: 421-475.

—— **& D. L. Olivier.** 1952. The taxonomy of *Ophiobolus graminis* Sacc. Trans. Brit. Mycol. Soc. **35**: 29-33.

—— **& E. Müller.** 1954. Die Gattungen der amerosporen Pyrenomyceten. Beitr. Kryptfl. Schweiz **11**(1): 1-454.

—— & ——. 1955. Über die Gattungen *Delitischia* Auersw., *Tricholdelitschia* Munk und *Cainia* nov. gen. Acta Bot. Neerland. **4**: 108-115.

—— & ——.1955a. Über die Gattungen *Selinia* Karst. und *Seliniella* nov. gen. und ihre phylogenetische Bedeutung. Acta Bot. Neerland. **4**: 116-125.

Atkinson, G. F. 1912. The morphology of *Zygorhynchus* and its relation to the Ascomycetes. Science **35**: 151.

——. 1915. Phylogeny and the relationships in the Ascomycetes. Ann. Missouri Bot. Gard. **2**: 315-376.

Bachus, E. J. & G. W. Keitt. 1940. Some nuclear phenomenon in *Venturia inaequalis*. Bull. Torrey Bot. Club **67**: 765-770.

Baranetsky, J. 1872. Entwicklungsgeschichte des *Gymnoascus reesii.* Bot. Z. **30**: 145.

Barker, D. T. 1905. The structure of the ascocarp in the genus *Monascus.* Rep. Brit. Assn. Adv. Sci. London **74**: 824-825.

Barr, M. E. 1956. The development of the ascocarp in *Phaeotrichum hystricinum*. Canad. J. Bot. **34**: 564-568.

——. 1958. Life history studies of *Mycosphaerella tassiana* and *M. typhae.* Mycologia **50**: 501-513.

Barron, G.I., R. F. Cain & J. C. Gilman. 1961. A revision of the genus *Petriella.* Canad. J. Bot. **39**: 837-845.

——, —— & ——. 1961a. The genus *Microascus.* Canad. J. Bot. **39**: 1609-1631.

Batra, L. R. 1959. A comparative morphological and physiological study of the species of *Dipodascus.* Mycologia **51**: 329-355.

Beadle, G. N. & E. L. Tatum. 1941. *Neurospora* I. Genetic control of biochemical reactions in *Neurospora.* Proc. Nat. Acad. Sci. **27**: 499.

—— & ——. 1945. *Neurospora* II. Methods of producing and detecting mutations concerned with nutritional requirements. Amer. J. Bot. **32**: 678-686.

Beeli, M. 1920. Note sur le genre *Meliola* Fr. Bull. Jard. Bot. Brussels **9**: 89-160.

Benjamin, C. R. 1955. Ascocarps of *Aspergillus* and *Penicillium.* Mycologia **47**: 669-687.

Benjamin, R. K. 1956. A new genus of the Gymnoascaceae with a review of the other genera. El Aliso **3**: 301-328.

—— **& L. Shanor.** 1950. Discovery of dioecism in *Laboulbenia formicarum*. Science **111**: 33-34.

—— & ——. 1950a. The development of male and female individuals in the dioecious species *Laboulbenia formicarum* Thaxter. Amer. J. Bot. **37**: 471-476.

—— & ——. 1951. Morphology of immature stages of *Euzodiomyces lathrobi* Thaxter and the taxonomic position of the genus *Euzodiomyces*. Amer. J. Bot. **38**: 555-560.

—— & ——. 1952. Sex of host specificity and position specificity of *Laboulbenia* on *Bembidion picipes*. Amer. J. Bot. **39**: 125-131.

Biggs, R. 1937. *Dipodascus uninucleatus*. Mycologia **29**: 34-44.

Bitancourt, A. A. & A. E. Jenkins. 1935. Areolate spot of citrus caused by *Leptosphaeria bondari.* Phytopathology **25**: 884-886.

—— & ——. 1936. Perfect stages of the sweet orange fruit scab fungus. Mycologia **28**: 489-492.

—— & ——. 1937. Sweet orange fruit scab caused by *Elsinoe australis*. J. Agr. Res. **54**: 1-18.

Blackman, V. H. & E. J. Welsford. 1912. The development of the perithecium of *Polystigma rubrum* D.C. Ann. Bot. **26**: 761-767.

Blumer, S. 1933. Die Erysipheen Mitteleuropas. Beitr. Kryptfl. Schweiz **7**(1): 1-483.

Boone, D. M., D. M. Kline & G. W. Keitt. 1957. *Venturia inaequalis* (Cke.) Wint. VIII. Pathogenicity of induced biochemical mutants. Amer. J. Bot. **44**: 791-796.

Booth, C. 1957. Studies of the Pyrenomycetes I. Four species of *Chaetosphaeria,* two with *Catenularia* conidia. II. *Melanopsamma pomiformis* and its *Stachybotrys* conidia. Comm. Mycol. Inst. Mycol. Pap. **68**: 1-27.

——. 1958. The genera *Chaetosphaeria* and *Thaxteria* in Britain. The Naturalist **1958**: 83-90.

Bose, S. K. 1961. Studies on *Massarina* Sacc., and related genera. Phytopathol. Z. **41**: 151-213.

Brandriff, H. 1936. The development of the ascocarp of *Acrospermum compressum.* Mycologia **28**: 228-235.

Brefeld, O. 1872-1908. Botanische Untersuchungen über Schimmelpilze. Vol. 1-14. Verlag Arthur Felix, Leipzig.

Brooks, F. T. 1910. The development of *Gnomonia erythrostoma* Pers. Ann. Bot. **24**: 585-605.

Bubak, F. 1909. Eine neue Krankheit der Luzerne in Oesterreich. Wien Landw. Zeit. **59**: 908-909.

Buisman, C. J. 1929. Onderzoek naar de oorzaak von de iepenziekte II. De Iepenziekte **1929**: 13-78.

——. 1932. *Ceratostomella ulmi,* de Geslachtelijka Vorm von *Graphium ulmi* Schwarz. Tijdschr. Plantenziekt. **38**: 1-8.

Cain, R. F. 1934. Studies of coprophilous Sphaeriales in Ontario. Univ. Toronto Press, Biol. Ser. No. 38, 136 pp.

——. 1948. *Myriogonium,* a new genus among simplified ascomycetes. Mycologia **30**: 158-167.

——. 1948a. Notes on seed-borne fungi. VI. *Sordaria.* Canad. J. Bot. **20**: 486-495.

——. 1950. Studies of coprophilous ascomycetes. I. *Gelasinospora.* Canad. J. Bot. **28**: 566-576.

——. 1956. Studies of soil fungi, *Saturnomyces,* a new genus of the Aspergillaceae. Canad. J. Bot. **34**: 135-141.

——. 1956a. Studies in the coprophilous ascomycetes. II. *Phaeotrichum,* a new cleisto-carpous genus in a new family, and its relationships. Canad. J. Bot. **34**: 675-687.

——. 1956b. Studies of coprophilous ascomycetes. IV. *Tripterospora,* a new cleistocarpous genus in a new familv. Canad. J. Bot. **34**: 699-710.

——. 1956c. *Synaptospora,* a new genus of amerosporous Ascohymeniales (Ascomycetes). Sydowia, Ser. II, Beihft **1**: 4-8.

——. 1961. Studies of soil fungi. III. New species of *Coniochaeta, Chaetomidium* and *Thielavia.* Canad. J. Bot. **39**: 1231 1239.

——. 1961a. Studies of coprophilous ascomycetes. VII. *Preussia.* Canad. J. Bot. **39**: 1635-1666.

——. 1962. Studies of coprophilous ascomycetes. VIII. New species of *Podospora.* Canad. J. Bot. **40**: 447-490.

—— **& L. K. Weresub.** 1957. Studies of coprophilous ascomycetes. V. *Sphaeronemella fimicola.* Canad. J. Bot. **35**: 119-131.

Cash, E. K. & A. M. Waterman. 1957. A new species of *Plagiostoma* associated with a leaf disease of hybrid aspens. Mycologia **49**: 756-769.

Cavara, F. 1893. Ueber einige parasitische Pilze aus dem Getreide. Zeitschr. Pflanzenkr. **3**: 16-26.

Cayley, D. M. 1932. Some observations on the life history of *Nectria galligena* Bres. Ann. Bot. 35: 79-92.

Chenantais, M. J. E. 1920. Sillon et pores germinatifs. Bull. Soc. Mycol. Fr. **26**: 29-33.

Chesters, C. G. C. 1938. Studies of British Pyrenomycetes. II. A comparative study of *Melanomma pulvis-pyrius* (Pers.) Fuckel, *Melanomma fuscidulum* Sacc. and *Thyridaria rubro-notata* (B. & Br.) Sacc. Trans. Brit. Mycol. Soc. **22**: 116-150.

Chilton. S. J. P., G. B. Lucas & C. W. Edgerton. 1945. Genetics of *Glomerella*. III. Crosses with a conidial strain. Amer. J. Bot. **32**: 549-554.

—— **& H. E. Wheeler.** 1949. Genetics of *Glomerella*. VI. Linkage. Amer. J. Bot. **36**: 270-273.

—— **&** ——. 1949a. Genetics of *Glomerella*. VII. Mutation and segregation in plus cultures. Amer. J. Bot. **36**: 717-721.

Chivers, A. H. 1915. A monograph of the genera *Chaetomium* and *Ascotricha*. Mem. Torrey Bot. Club **14**(3): 155-240.

Claussen, P. 1905. Zur Entwicklungsgeschichte der Ascomyceten *Boudiera*. Bot. Z. **63**: 1-27.

——. 1906. Über neuere Arbeiten z. Entwicklungsgeschichte der Ascomyceten. Ber. Deut. Bot. Ges. **24**: 11-38.

——. 1907. Zur Kenntnis der Kernverhältnisse von *Pyronema confluens*. Ber. Deut. Bot. Ges. **25**: 586-590.

——. 1912. Zur Entwicklungsgeschichte der Ascomyceten *Pyronema confluens*. Zeitschr. Bot. **4**: 1-64.

Clements, F. E. & C. L. Shear. 1931. The genera of fungi. 496 pp. H. W. Wilson Co., New York.

Clum, F. M. 1955. A new genus in the Aspergillaceae. Mycologia **47**: 899-901.

Colson, B. 1938. The cytology and development of *Phyllactinia corylea*. Ann. Bot. n.s. **2**: 381-401.

Conners, I. L. 1959. Species of *Leptothyrium* and *Kabatia* on *Lonicera*. Canad. J. Bot. **37**: 419-401.

——. 1959a. Further observations on *Leptothyrium* and *Kabatia* on *Lonicera*. Canad J. Bot. **37**: 1131.

Cook, M. T. 1923. The life history of *Nectria ipomoeae*. Mycologia **15**: 233-235.

Cooper, J. R. 1917. Studies of the etiology and control of blister canker on apple trees. Univ. Nebraska Exp. Sta. Bull. **12**: 1-117.

Conant, N. F., D. S. Martin, D. T. Smith, R. D. Baker & J. L. Callaway. 1954. Manual of clinical mycology. 2nd ed. W. B. Saunders, Philadelphia.

Corbaz, R. 1957. Recherches sur les genre *Didymella*. Phytopath. Z. **28**: 275-414.

Corlett, M. 1963. The developmental morphology of two species of *Microascus*. Canad. J. Bot. **41**: 837-845.

——. 1965. Perithecium development in *Chaetomium trigonosporum*. Canad. J. Bot. **44**: 155-162.

——. 1966. Developmental studies in the Microascaceae. Canad. J. Bot. **44**: 79-88.

Dance, B. W. 1959. A cultural connection between *Venturia tremulae* Adrh. and its imperfect stage in Ontario. Canad. J. Bot. **37**: 1139-1140.

——. 1961. Leaf spot and shoot blight of poplar (section *Tacamahaca* Spach.) caused by *Venturia populina* (Vuill.) Fabric. Canad. J. Bot. **39**: 875-890.

Dandero, J. B. 1907. Notes on the black knot of plum. Rep. Mich. Acad. Sci. **9**: 74-75.

Dangeard, P. A. 1907. L'origin du perithece chez les ascomycetes. Le Botaniste **10**: 1-385.

Darker, G. D. 1963. A new genus of the Lophiaceae. Canad. J. Bot. **41**: 1383-1399.

DeBary, A. 1881. Untersuchungen über die Peronosporeen und Saprolegnieen und die Grundlagen eines natürlichen Systems der Pilze. *In* DeBary, A. & M. Woronin. Beitr. Morph. Physiol. Pilze **4**: 1-145.

——. 1884. Vergleichende Morphologie und Biologie der Pilze, Mycetozoen und Bacterien. Wilhlem Engelmann, Leipzig.

——. 1887. Comparative Morphology and Biology of the Fungi, Mycetozoa, and Bacteria. Clarendon Press, Oxford.

Defago, G. 1942. Second contribution a la connaissance des Valsées von Höhnel. Phytopath. Z. **14**: 103-149.

DeLamater, E. D. 1937. Crozier formation in the Gymnoascaceae: A preliminary note. Mycologia **29**: 187-198.

——**, S. Yaverbaum & L. Schwartz.** 1953. The nuclear cytology of *Eremascus albus*. Amer. J. Bot. **40**: 475-492.

Diedicke, H. 1904. Ueber den Zusammenhang zwischen *Pleospora*- und *Helminthosporium*-Arten. II. Cęntrbl. Bact. Parsit. Infect., Abt. 2, **11**: 52-59.

Diehl, W. W. 1950. *Balansia* and the Balansiae in America. U. S. Dept. Agr. Mon. **4**: 1-82.

Dimock, A. W. 1937. Observations on sexual relations in *Hypomyces*. Mycologia **29**: 116-137.

Dingley, J. N. 1951. The Hypocreales of New Zealand. I. Trans. Roy. Soc. N. Z. **79**: 55-61.

——. 1951a. The Hypocreales of New Zealand. II. The genus *Nectria*. Trans. Roy. Soc. N. Z. **79**: 177-202.

——. 1952. The Hypocreales of New Zealand. III. The genus *Hypocrea*. Trans. Roy. Soc. N. Z. **79**: 323-337.

——. 1952a. The Hypocreales of New Zealand. IV. The genera *Calonectria, Giberella* and *Thyronectria*. Trans. Roy. Soc. N. Z. **79**: 403-411.

——. 1952b. The Hypocreales of New Zealand. V. The genera *Cordyceps* and *Torrubiella*. Trans. Roy. Soc. N. Z. **81**: 329-343.

——. 1954. The Hypocreales of New Zealand. VI. The genera *Hypocrella, Barya, Claviceps* and *Podonectria*. Trans. Roy. Soc. N. Z. **81**: 480-499.

——. 1956. The Hypocreales of New Zealand. VII. A revision of the records and species in the Hypocreaceae. Trans. Roy. Soc. N. Z. **83**: 643-662.

——. 1957. Life history studies of New Zealand species of *Nectria*. Trans. Roy. Soc. N. Z. **84**: 467-477.

——. 1957a. Life history studies in the genus *Hypocrea* Fr. Trans. Roy. Soc. N. Z. **84**: 689-693.

Dodge, B. O. 1927. Nuclear phenomenon associated with heterothallism and homothallism in the ascomycete *Neurospora*. J. Agr. Res. **35**: 289-305.

——. 1928. Spore formation in asci with fewer than eight spores. Mycologia **20**: 18-21.

——. 1928a. Production of fertile hybrids in the ascomycete *Neurospora*. J. Agr. Res. **36**: 1-14.

——. 1930. Breeding albinistic strains of the *Monilia* bread mold. Mycologia **22**: 9-38.

——. 1931. Inheritance of the albinistic non-conidial characters in interspecific hybrids of *Neurospora*. Mycologia **23**: 1-47.

——. 1935. The mechanics of sexual reproduction in *Neurospora*. Mycologia **27**: 418-438.

Doguet, G. 1955. Etude du développement du *Chaetoceratostoma longirostre* Farrow. Rev. Mycol. Suppl. Col. **20**: 132-143.

——. 1955a. Le genre "*Melanospora*," biologie, morphologie développement, systematique. Le Botaniste, **39**(I-VI): 1-313.

——. 1956. Le genre *Thielavia* Zopf, Rev. Mycol. Suppl. Col. **21**: 1-21.

——. 1957. Organogénie du *Creopus spinulosus* (Fuck.) Moravec. Organogénie de quelque Hypocreales du même type. Bull. Soc. Mycol. Fr. **73**: 144-164.

——. 1957a. Organogénie du *Microascus stysanophorus* (Matt.) Curzi. Bull. Soc. Mycol. Fr. **73**: 105-178.

——. 1960. Morphologie, organogénie et evolution nucleare de l'*Epichloe typhina*. La place des Clavicipitaceae dans la classification. Bull. Soc. Mycol. Fr. **76**: 171-203.

——. 1960a. Etude du *Melogramma spiniferum* (Wallr.) de Notaris, Pyrenomycète Ascohymenié, annellasce, bituniqué. Rev. Mycol. **35**: 13-36.

——. 1961. Recherches sur le *Wawelia regia*, morphologie et organogénie. Bull. Soc. Mycol. Fr. **77**: 197-218.

Doidge, E. M. 1942. A revision of the South African Microthyriaceae. Bothalia **4**(2): 273-420.

—— **& H. Sydow.** 1927. The South African species of the Meliolaceae. Bothalia **2**: 424-472.

Dowding, E. S. & E. H. Gowan. 1942. The migration of fungal nuclei in an electric field. Canad. J. Bot. **20**: 92-100.

Drechsler, C. 1923. Some graminicolous species of *Helminthosporium*. J. Agr. Res. **24**: 641-740.

——. 1925. Leafspot of maize caused by *Ophiobolus heterostrophus* n. sp., the ascigerous stage of *Helminthosporium* n. sp.. exhibiting bipolar germination. J. Agr. Res. **31**: 701-726.

——. 1928. Zonate eyespot of grasses caused by *Helminthosporium giganteum*. J. Agr. Res. **37**: 473-492.

——. 1934. Phytopathologic and taxonomic aspects of *Ophiobolus, Pyrenophora, Helminthosporium* and the genus *Cochliobolus*. Phytopathology **24**: 953-983.

Edgerton, C.W. 1914. Plus and minus strains in the genus *Glomerella*. Amer. J. Bot. **1**: 244-254.

——, **S. J. P. Chilton & G. B. Lucas.** 1945. Genetics of *Glomerella*. II. Fertilization between strains. Amer. J. Bot. **32**: 115-118.

Ehrlich, J. 1934. The beech bark disease. A *Nectria* disease on *Fagus* following *Cryptococcus fagi* (Baer.). Canad. J. Res. **10**: 593-692.

El-Ani, A. S. 1954. The genetics of sex in *Hypomyces solani* f. *cucurbitae*. Amer. J. Bot. **41**: 110-113.

——. 1956. Ascus development and nuclear behavior in *Hypomyces solani* f. *cucurbitae*. Amer. J. Bot. **43**: 769-777.

Elliott, J. 1925. A cytological study of *Ceratostomella fimbriata* (E. & H.) Elliott. Phytopathology **15**: 417-422.

Ellis, J. B. & B. M. Everhart. 1892. The North American Pyrenomycetes. Pub. by authors, Newfield, New Jersey.

Emmons, C. W. 1935. The ascocarps in species of *Penicillium*. Mycologia **27**: 129-149.

—— **& B. O. Dodge.** 1931. The ascosporic stage of species of *Scopulariopsis*. Mycologia **23**: 313-331.

Farr, M. L. 1963. The systematic position of some *Dimeriella* species and associated fungi on Pinaceae. Mycologia **55**: 226-246.

Faull, J. H. 1912. The cytology of *Laboulbenia chaetophora* and *G. gyrinidarum*. Ann. Bot. **26**: 325-355.

Faurel, L. & G. Schotter. 1959. Un Champignon ascomycète inédit, type d'une nouvelle famille de position systematique critique. Compt. Rend. Acad. Sci. Paris. **249**: 150-152.

Fisch, C. 1882. Beiträge zur Entwicklungsgeschichte einiger Ascomyceten. Bot. Z. **40**: 851-869.

Fitzpatrick, H. M. 1923. Monograph of the Nitschkieae. Mycologia **15**: 23-44 & 45-67.

——. 1942. Revisionary studies in the Coryneliaceae. Mycologia **34**: 464-488.

——. 1942a. Revisionary studies in the Coryneliaceae. II. The genus *Caliciopsis*. Mycologia **34**: 489-514.

Frank, B. 1883. Ueber einige neue oder weniger bekannte Pflanzenkrankheiten. II. Ber. Deut. Bot. Ges. **1**: 58-63.

——. 1886. Über *Gnomonia erythrostoma* etc. Ber. Deut. Bot. Ges. **4**: 200-205.

Fraser, L. 1935. An investigation of the sooty molds of New South Wales. III. Proc. Linn. Soc. N. S. Wales **60**: 97-118.

Fries, E. 1821-1823. Systema Mycologicum **1**: 1-529. 1821; **2**: 1-621. 1823. Grypswalddeae.

——. 1846-1849. Summa vegetabilium Scandinaviae. 1-258. 1846; 259-572. 1849. Bonnier Holmiae et Lipsiae.

Fuckel, L. 1869-70. Symbolae Mycologicae et Appendices. i-iii. pp. 1-459. Wiesbaden.

Funk, A. 1964. *Botryosphaeria tsugae* n. sp. causing a dieback of western hemlock in British Columbia. Canad. J. Bot. **42**: 769-775.

——. 1965. A new parasite of spruce from British Columbia. Canad. J. Bot. **43**: 45-48.

—— **& A. K. Parker.** 1966. *Scirrhia pini* n. sp., the perfect stage of *Dothiostroma pini* Hulberry. Canad. J. Bot. **44**: 1171-1176.

Gaillard, A. 1892. Le genre *Meliola*. Bull. Soc. Myçol. Fr. **8**: 76-188.

Gaeumann, E. 1949. Die Pilze. 392 pp. Basel.

Gilman, J. C. & L. H. Tiffany. 1951. Iowa Ascomycetes. I. The Xylariaceae. Iowa J. Sci. **26**: 455-482.

Goidanich, G. 1936. Id genera di ascomyceti *Grossmania* G. Goid. Boll. staz. patol. vegetale, n. s. **16**: 26-60.

Graff, P. W. 1932. The morphological and cytological development of *Meliola circinans.* Bull. Torrey Bot. Club **59**: 241-266.

Graham, J. M. & E. S. Luttrell. 1961. Species of *Leptosphaerulina* on forage plants. Phytopathology **51**: 680-693.

Graves, A. H. 1923. The *Melanconis* disease of the butternut (*Juglans cinerea*). Phytopathology **13**: 411-435.

Gremmen, J. 1951. En blad- en twigziekte van Populieren door *Venturia tremulae* en *Venturia populina.* Korte Medded. No. 27 bosbouwproefst. 236-242.

Griffiths, D. 1901. The North American Sordariaceae. Mem. Torrey Bot. Club **11**: 1-134.

Grossenbacher, J. G. & B. M. Duggar. 1911. A contribution to the life-history, parasitism and biology of *Botryosphaeria ribis.* N. Y. Exp. Sta. Techn. Bull. **18**: 114-188.

Guillermond, A. 1940. Sexuality, developmental cycle and phylogeny of yeasts. Bot. Rev. **6**: 1-24.

Gwynne-Vaughan, H. C. I. 1922. Fungi. Ascomycetes, Ustilaginales, Uredinales. Cambridge Univ. Press.

—— **& Q. E. Broadhead.** 1936. Contributions to the study of *Ceratostomella fimbriata.* Ann. Bot. **50**: 747-758.

Hanlin, R. T. 1961. Studies in the genus *Nectria.* II. Morphology of *N. gliocladiodes*. Amer. J. Bot. **48**: 900-908.

——. 1962. Morphology of *Hypomyces lactifluorum*. Bot. Gaz. **124**: 395-404.

——. 1963. Morphology of *Neuronectria peziza.* Amer. J. Bot. **50**: 56-66.

——. 1964. Morphology of *Hypomyces tricothecoides.* Amer. J. Bot. **51**: 201-208.

——. 1965. Morphology of *Hypocrea schweinitzii*. Amer. J. Bot. **52**: 570-579.

Hansen, H. N. 1938. The dual phenomenon in imperfect fungi. Mycologia **30**: 442-455.

—— **& W. C. Snyder.** 1943. The dual phenomenon and sex in *Hypomyces solani* f. *cucurbitae.* Amer. J. Bot. **30**: 419-422.

Hansford, C. G. 1946. The foliaceous ascomycetes, their parasites and associated fungi. Imp. Mycol. Inst., Kew Mycol. Pap. **15**: 1-240.

Harper, R. A. 1895. Die Entwicklung des Peritheciums bei *Sphaerotheca castagnei.* Ber. Deut. Bot. Ges. **13**: 475-481.

——. 1905. Sexual reproduction and the organization of the nucleus in certain mildews. Carnegie Inst. Washington Publ. **37**: 1-105.

Hemmi, T. 1916. On the die-back disease of *Paulownia tomentosa* caused by a new species of *Valsa*. Bot. Mag. **30**: 304-415.

Hess, H. & E. Müller. 1951. Zur Entwicklungsgeschichte von *Dothidella insculpta* (Wallr.) Theiss. & Syd. Ber. Schweiz Bot. Ges. **61**: 5-34.

Higgins, B. B. 1920. Morphology and life history of some Ascomycetes, with special reference to the presence and function of spermatia. Amer. J. Bot. **7**: 435-444.

——. 1929. Morphology and life history of some Ascomycetes with special reference to the presence and function of spermatia. II. Amer. J. Bot. **16**: 287-296.

——. 1936. Morphology and life history of some Ascomycetes with special reference to the presence and function of spermatia. III. Amer. J. Bot. **23**: 598-602.

Hilitzer, A. 1929. Monografická studie o ceshych druzich radu Hysteriales a o sypavkách jimi pusobenych. (Étude monographique sur les espèces de l'ordre Hysteriales, trouvées en Boheme, et sur les epiphytes qui en ont causées). Vedecke spisy vydávané Ceskoslocenskou akademii zemeldelskou **3**: 1-162.

Hirsch, H. E. F. 1949. The cytogenetics of sex in *Hypomyces solani* f. *cucurbitae.* Amer. J. Bot. **36**: 113-121.

Höhnel, F. von. 1907. *Wettsteinina* n. g. (Pseudosphaeriaceae). Fragmente zur Mykologie 128. Sitz. Akad. Wiss. Wien. **116**: 126-129.

——. 1917. System der Diaportheen. Ber. Deut. Bot. Ges. **35**: 631-638.

——. 1918. Rehm: Ascomycetes exs. Fasc. 56 & 57. Ann. Myc. **16**: 209-224.

——. 1919. Über den Zusammenhang von *Meliola* mit den Microthyriaceen. Ber. Deut. Bot. Ges. **36**: 471-473.

Holm, L. 1948. Taxonomical notes on Ascomycetes. I. The Swedish species of the genus *Ophiobolus* Riese sensu Sacc. Sartyr. Svensk Bot. Tidskr. **42**: 337-347.

——. 1952. Taxonomical notes on Ascomycetes. II. The herbicolous Swedish species of the genus *Leptosphaeria* Ces. & de Not. Sartyr. Svensk. Bot. Tidskr. **46**: 18-46.

——. 1957. Édutes taxonomiques sur les Pleosporacees. Symb. Bot. Uppsaliensis **14**: 1-188.

——. 1967. Taxonomic notes on Ascomycetes. V. On *Sphaeria parmeliarum* Phill & Plowr. and the genus *Cucurbidothis* Petr. Svensk Bot. Tidskr. **61**: 449-456.

——. 1968. Taxonomic notes on Ascomycetes. VI. On the genus *Gibberidea* Fuck. and some alleged relatives. Svensk Bot. Tidskr. **62**: 219-242.

Holterman, C. 1898. Mykologische Untersuchungen aus den Tropen. pp. 1-122. Berlin.

Houlihan, M. B., G. W. Beadle, & H. G. Calhoun. 1949. Linkage studies with biochemical mutants of *Neurospora crassa.* Genetics **34**: 493-507.

Hrushovetz, S. B. 1956. Cytological studies of the ascus development of *Cochliobolus sativus.* Canad. J. Bot. **34**: 641-651.

Hubbes, M. 1962. Inhibition of *Hypoxylon pruinatum* by pyrocatechol isolates from bark of aspen. Science **136**: 156.

——. 1964. New facts on host parasitic relationships in the *Hypoxylon* canker of aspen. Canad. J. Bot. **42**: 1489-1494.

Hughes, S. J. 1952. *Sirodesmium granulosum* and *Torula diversa.* The Naturalist **1952**: 93-98.

——. 1953. Some foliicolous hyphomycetes. Canad. J. Bot. **31**: 565-576.

——. 1965. New Zealand fungi. 3. *Catenularia* Grove. N. Z. J. Bot. **3**: 136-150.

——. 1966. New Zealand fungi. 6. N. Z. J. Bot. **4**: 77-85.

Hunt, J. 1956. Taxonomy of the genus *Ceratocystis.* Lloydia **19**: 1-59.

Huttig, W. 1935. Die Sexualität bei *Glomerella lycopersici* und ihre Vererbung. Biol. Zentralbl. **55**: 74-83.

Ingold, C. T. 1954. The ascogenous hyphae in *Daldinia.* Trans. Brit. Mycol. Soc. 108-110.

——. 1965. Spore Liberation. Oxford Press.

Jackson, H. S. 1947. *Trichomonas*, a new genus among simple ascomycetes. Mycologia **39**: 709-715.

Jaczewski, F. A. de. 1894. Essai de classification naturelle de Pyrenomycetes. Bull. Soc. Mycol. Fr. **10**: 13-48.

Jeffers, W. F. 1940. Studies on *Caryospora putaminum*. Mycologia **32**: 550-569.

Jenkins, A. 1931. Scab of *Canavalia* caused by *Elsinoe canavaliae*. J. Agr. Res. **42**: 1012.

——. 1931a. Lima-bean scab caused by *Elsinoe*. J. Agr. Res. **42**: 13-23.

——. 1931b. Development of the citrus-scab organism. J. Agr. Res. **42**: 545-558.

——. 1932. *Elsinoe* on apple and pear. J. Agr. Res. **44**: 689-700.

——. 1938. Rose anthracnose. Amer. Rose Annual **126**: 141.

—— **& A. A. Bitancourt.** 1957. Studies in the Myriangiales. VII. Elsinoaceae on evergreen *Evonymus*, rose and English ivy. Mycologia **49**: 95-101.

Jenkins, W. A. 1934. The development of *Cordyceps agariciformia*. Mycologia **26**: 220-243.

Johnson, T. W. & F. K. Sparrow. 1961. Fungi in Oceans and Estuaries. 669 pp. J. Cramer, Weinheim.

Jones, S. C. 1926. The development of the perithecium of *Ophiobolus graminis* Sacc. Ann. Bot. **40**: 607-629.

Keitt, G. W. & W. H. Langford. 1941. *Venturia inaequalis* (Cke.) Wint. I. A ground work for genetic studies. Amer. J. Bot. **28**: 805-820.

——**, M. H. Langford & J. R. Shay.** 1943. *Venturia inaequalis* (Cke.) Wint. II. Genetic studies in pathogenicity and certain mutant characters. Amer. J. Bot. **30**: 491-500.

Kern, H. 1957. Untersuchungen über die Umgrenzung der Arten in der Ascomycetengattung *Leucostoma*. Phytopath. Z. **30**: 149-180.

Killian, C. 1917. Über die Sexualität von *Venturia inaequalis* (Cke.) Ad. Z. Bot. **9**: 353-397.

——. 1919. Sur la sexualité de l'Ergot du seigle, le *Claviceps purpurea* (Tulasne). Bull. Soc. Mycol. Fr. **35**: 182-197.

——. 1920. Le développement de *Dothidella ulmi*. Rev. Gén. Bot. **32**: 534-551.

——. 1922. Le développement du *Stigmatea robertiani* Fries. Rev. Gén. Bot. **34**: 577-588, pl. 14-17.

Kirschstein, W. 1933. Remarks on a collection of British species of *Rosellinia* and a redistribution of the species in that genus. Trans. Brit. Mycol. Soc. **18**: 302-307.

Klebahn, H. 1918. Haupt- und Nebenfruchtformen der Askomyzeten. 395 pp. Gebrüder Borntraeger, Leipzig.

Kline, G. M., D. M. Boone & G. W. Keitt. 1957. *Venturia inaequalis*. XIV. Nutritional control of pathogenicity of certain induced biochemical mutants. Amer. J. Bot. **44**: 797-803.

Kobayashi, T. 1970. Taxonomic studies of Japanese Diaporthaceae. with special reference to their life-histories. Bull. Govt. For. Exp. Sta. Tokyo **226**: 1-242.

Kobayasi, Y. 1941. The genus *Cordyceps* and its allies. Sci. Rep. Tokyo Bun Daigaku **5**: 253-260.

Kowalski, D. T. 1964. The development and cytology of *Pycnidiophora dispersa*. Amer. J. Bot. **51**: 1076-1082.

——. 1965. The development and cytology of *Melanospora tiffanii*. Mycologia **57**: 279-290.

——. 1965a. The development and cytology of *Preussia typharum*. Bot. Gaz. **126**: 123-130.

Kramer, C. L. 1958. A new genus of the Protomycetaceae. Mycologia **50**: 916-926.

Kuehn, H. M. 1958. A prelininary survey of the Gymnoascaceae. I. Mycologia **50**: 417-440.

Kuyper, H. P. 1905. Die Peritheciumentwicklung von *Monascus bakeri* Dangeard, sowie die systematische Stellung dieser Pilze. Ann. Mycol. **3**: 32.

Leben, C. C. & J. R. Shav. 1948. *Venturia inaequalis* (Cke.) Wint. IV. Further studies on the inheritance of pathogenicity. Amer. J. Bot. **35**: 334-336

Leopold, U. 1950. Die Vererbung von Homothallie und Heterothallie bei *Schizosaccharomyces pombe*. Compt. Rend. Lab. Carlsberg ser. Phys. **24**: 281-480.

Likhite, V. 1926. Recherches sur le développement et la biolgie de quelques Ascomycètes. Rev. Gén. Bot. **38**: 5-30.

Lindau, G. 1897. Pyrenomycetineae. *In* Engler & Prantl. "Die natürlichen Pflanzenfamilien. I." **1**: 321-491.

Lindegren, C. C. 1932. The genetics of *Neurospora*. I. Inheritance of response to heat treatment. Bull. Torrey Bot. Club **59**: 85-102.

——. 1932a. The genetics of *Neurospora*. II. Segregation of sex characters in the asci of *N. crassa, N. sitophila* and *N. tetrasperma*. Bull. Torrey Bot. Club **59**: 119-138.

——. 1933. The genetics of *Neurospora*. III. Purebred stocks and crossing over in *N. crassa*. Bull. Torrey Bot. Club **60**: 133-154.

——. 1934. The genetics of *Neurospora*. IV. The inheritance of tan versus normal. Amer. J. Bot. **21**: 55-65

——. 1934a. The genetics of *Neurospora*. V. Self-sterile bisexual heterokaryons. J. Genetics **28**: 425-435.

——. 1934b. The genetics of *Neurospora*. VI. Bisexual and karyotic ascospores from *N. crassa*. Genetics **16**: 315-320.

——. 1936. The structure of the sex chromosomes of *Neurospora crassa*. J. Heredity **27**: 251-259.

——. 1936a. A six point map of the sex chromosomes of *Neurospora crassa*. J. Genetics **32**: 243-256.

——. 1936b. Heterokaryosis and hormones in *Neurospora* Amer. Nat. **76**: 404-405.

——. 1942. The use of fungi in modern genetical analysis. Iowa J. Sci. **16**: 271-290.

——. 1949. Genetics of Fungi. Ann. Rev. Microbiol. **1949**: 47-90.

——. 1949a. The yeast cell, its genetics and cytology. Educational Publishers Inc., St. Louis.

Lodder, J. & N. J. W. Kreger-van Rij. 1952. The yeasts, a taxonomic study. Pp. 1-713. North Holland Publ. Col., N. Y.

Loeffler, W. 1957. Untersuchungen über die Ascomyceten-Gattung *Dothidea* Fr. Phytopath. Z. **30**: 349-386.

—— **& E. Müller.** 1961. Über *Valsonectria reticulata*, eine neue Art einer wenig bekannten Ascomyceten-Gattung. Ber. Schweiz Bot. Ges. **71**: 409-415.

Lohman, M. L. 1931. A study of *Glonium parvulum* in culture. Pap. Mich. Acad. Sci. **13**: 141-157.

——. 1932. Three new species of *Mytilidion* in the proposed subgenus *Lophiopsis*. Mycologia **24**: 477-485.

——. 1932a. The comparative morphology of germinating ascospores in certain species of Hysteriaceae. Pap. Mich. Acad. Sci. **15**: 97-109.

——. 1933. *Septonema toruloideum*; a stage of *Mytilidion scolecosporum*. Mycologia **25**: 34-42.

——. 1933a. Hysteriales: Life-histories of certain species. Pap. Mich. Acad. Sci. **17**: 229-288.

——. 1934. *Lophiosphaeria (Glonium) velata*, with a critical study of its *Septonema multiplex* stage. Amer. J. Bot. **21**: 314-327.

——. 1934a. A cultural and taxonomic study of *Hysterium hyalinum*. Pap. Mich. Acad. Sci. **19**: 133-140.

——. 1937. Studies in the genus *Glonium* as represented in the Southeast. Bull. Torrey Bot. Club **64**: 57-73.

——. 1938. Observations on a species of *Bulliardiella*. Pap. Mich. Acad. Sci. **23**: 155-162.

—— **& A. J. Watson.** 1953. Identity and host relations of *Nectria* species associated with diseases of hardwoods in the eastern states. Lloydia **6**: 77-108.

Luc, M. 1952. Structure et développement de deux Dothidealés: *Systremma natans* (Tode) Th. et Syd. et *Bertia moriformis* (Tode) de Not. Bull. Soc. Mycol. Fr. **68**: 149-164.

Lucas, G. B. 1946. Genetics of *Glomerella* IV. Nuclear phenomenon in the ascus. Amer. J. Bot. **33**: 802-806.

——**, S. J. P. Chilton & C. W. Edgerton.** 1944. Genetics of *Glomerella*. I. Studies on the behavior of certain strains. Amer. J. Bot. **31**: 233-239.

Lucas, M. T. 1963. Culture studies on Portuguese species of *Leptosphaeria*. Trans. Brit. Mycol. Soc. **46**: 361-367.

——. 1965. Two species of *Pleospora* on succulents and their life history. Trans. Brit. Mycol. Soc. **48**: 415-416.

—— **& J. Webster.** 1964. Conidia of *Pleospora scirpicola* and *P. valesiaca*. Trans. Brit. Mycol. Soc. **47**: 247-256.

—— **& ——.** 1967. Conidial states of British species of *Leptosphaeria*. Trans. Brit. Mycol. Soc. **50**: 85-121.

Luttrell, E. S. 1940. *Morenoella quercina*, cause of a leaf spot of oaks. Mycologia **32**: 652-666.

——. 1944. The morphology of *Sphaerostilbe aurantiicola* (B. & Br.) Petch. Bull. Torrey Bot. Club **71**: 599-619.

——. 1944a. The morphology of *Myiocopron smilacis* (de Not.) Sacc. Amer. J. Bot. **31**: 640-649.

——. 1946. The genus *Stomiopeltis* (Hemisphaeriaceae). Mycologia **38**: 565-586.

——. 1951. The morphology of *Dothidea collecta*. Amer. J. Bot. **38**: 460-471.

——. 1951a. Taxonomy of the pyrenomycetes. Univ. Missouri Stud. **24**(3): 1-120.

——. 1953. Development of the ascocarp in *Glonium stellatum*. Amer. J. Bot. **40**: 626-633.

——. 1955. The ascostromatic Ascomycetes. Mycologia **47**: 511-532.

——. 1958. The perfect stage of *Helminthosporium turcicum*. Phytopathology **48**: 281-287.

——. 1960. The morphology of an undescribed species of *Dothiora*. Mycologia **52**: 64-79.

——. 1963. A *Trichometasphaeria* perfect stage for a *Helminthosporium* causing a leaf blight of *Dactyloctenium*. Phytopathology **53**: 281-285.

——. 1963a. Taxonomic criteria in *Helminthosporium*. Mycologia **55**: 643-674.

——. 1964. Systematics of *Helminthosporium* and related genera. Mycologia **56**: 119-132.

——. 1964a. Morphology of *Trichometasphaeria turcica*. Amer. J. Bot. **51**: 213-219.

——. 1965. Paraphysoids, pseudoparaphyses and apical paraphyses. Trans. Brit. Mycol. Soc. **48**: 135-144.

Mains, E. B. 1934. The genera *Cordyceps* and *Ophiocordyceps* in Michigan. Proc. Amer. Phil. Soc. **74**: 263-271.

——. 1937. A new species of *Cordyceps* with notes concerning other species. Mycologia **29**: 674-677.

——. 1939. *Cordyceps* from the mountains of North Carolina and Tennessee. J. Elisha Mitchell Sci. Soc. **55**: 117-129.

——. 1940. *Cordyceps* species from British Honduras. Mycologia **32**: 16-22.

——. 1941. *Cordyceps stylophora* and *Cordyceps ravenelii*. Mycologia **33**: 611-617.

——. 1947. New and interesting species of *Cordyceps*. Mycologia **39**: 535-545.

——. 1949. *Cordyceps bicephala* Berk. & Curt. and *C. australis* (Speg.) Sacc. Bull. Torrey Bot. Club **76**: 24-30.

——. 1950. Entomogenous species of *Akanthomyces, Hymenostilbe* and *Insecticola* in North America. Mycologia **42**: 566-589.

——. 1951. Entomogenous species of *Hirsutella, Tilamachlidium* and *Synnematium*. Mycologia **43**: 691-717.

——. 1954. Species of *Cordyceps* on spiders. Bull. Torrey Bot. Club **81**: 492-500.

——. 1955. Some entomogenous species of *Isaria*. Pap. Mich. Acad. Sci. **40**: 23-32.

——. 1957. Information concerning species of *Cordyceps* and *Ophionectria* in the Lloyd Herbarium. Lloydia **20**: 219-227.

——. 1957a. Species of *Cordyceps* parasitic on *Elaphomyces*. Bull Torrey Bot. Club **84**: 243-251.

——. 1958. North American entomogenous species of *Cordyceps*. Mycologia **50**: 169-222.

——. 1959. *Cordyceps* species. Bull. Torrey Bot. Club **86**: 46-58.

——. 1959a. North American species of *Aschersonia* parasitic on Aleyrodidae. J. Insect Path. **1**: 43-47.

Markert, C. L. 1949. Sexuality in the fungus *Glomerella*. Amer. Nat. **83**: 227-231.

Mason, E W. 1941. Annotated account of fungi received at the Mycological Institute. List II, Fasc. 3, Special pt. Commonw. Mycol. Inst. Mycol. Pap. **5**: 101-149.

Massee, G. 1895. A revision of the genus *Cordyceps*. Ann. Bot. **9**: 1-44.

Matruchot, L. & C. Dassonville. 1901 *Eidamella spinosa*, dermatophyte produisant des peritheces. Bull. Soc. Mycol. Fr. **17**: 123-132.

McClintock, B. 1945. *Neurospora* I. Preliminary observations of the chromosomes of *Neurospora crassa*. Amer. J. Bot. **32**: 671-678.

McCormack, H. W. 1936. The morphology and development of *Caliciopsis pinea*. Mycologia **28**: 188-196.

McGahen, J. W. & H. E. Wheeler. 1951. *Glomerella* IX. Perithecial development and plasmogamy. Amer. J. Bot. **38**: 610-617.

Miles, L. E. 1921. Leaf spots of elm. Bot. Gaz. **71**: 161-196.

Miller, J. H. 1928. Biologic studies in the Sphaeriales. Mycologia **20**: 187-213; 305-339.

——. 1930. British Xylariaceae. Trans. Brit. Mycol. Soc. **15**: 134-154.

——. 1932. British Xylariaceae. III. Trans. Brit. Mycol. Soc. **17**: 136-146.

——. 1938. Studies in the development of two *Myriangium* species and the systematic position of the Myriangiales. Mycologia **30**: 158-181.

——. 1940. The genus *Myriangium* in North America. Mycologia **32**: 587-600.

——. 1949. A revision of the classification of the ascomycetes with special emphasis on the pyrenomycetes. Mycologia **41**: 99-127.

——. 1951. Studies in the Phyllachoraceae I. *Phyllachora ambrosiae* (Berk. & Curt.) Sacc. Amer. J. Bot. **38**: 830-834.

—— 1954. Studies in the Phyllachoraceae. II. *Phyllachora lespedezae*. Amer. J. Bot. **41**: 825-828.

——. 1961. A monograph of the world species of *Hypoxylon*. Univ. Ga. Press, Athens, 158 pp.

—— **& M. G. Burton.** 1943. Study of *Bagnisiopsis* species on the Melastomaceae. Mycologia **35**: 312-334.

—— **& F. A. Wolf.** 1936. A leaf spot of honey locust caused by a new species of *Linospora*. Mycologia **28**: 171-180.

Mittmann, G. 1932. Kulturversuche mit Ensporstämmen und zytologische Untersuchungen in der Gattung *Cerastostomella*. Jahr. Wiss. Bot. **77**: 185-219.

Mix, A. J. 1935. The life history of *Taphrina deformans*. Phytopathology **25**: 41-66.

——. 1949. A monograph of the genus *Taphrina*. Sci. Bull. Univ. Kansas **33**: 1-167.

Morgan-Jones, J. F. 1953. Morpho-cytological studies of the genus *Gnomonia* I. *Gnomonia intermedia* Rehm; its development in pure culture. Svensk Bot. Tidskr. **47**: 284-308.

Mounce, I. & W. W. Diehl. 1934. A new *Ophiobolus* on eel grass. Canad. J. Res. **11**: 242-246.

Müller, E. 1950. Die Schweizerischen Arten der Gattung *Leptosphaeria* und ihre Verwandten. Sydowia **4**: 185-319.

——. 1951. Über die Entwicklung von *Pleospora gaeumannii* nov. spec. Ber. Schweiz. Bot. Ges. **61**: 165-174.

——. 1952. Die Schweizerischen Arten der Gattung *Ophiobolus* Riess. Ber. Schweiz. Bot. Ges. **62**: 307-339.

——. 1954. Beiträge zur Kenntnis der Gattung *Gibbera* Fr. emend Petr. Sydowia **8**: 60-73.

——. 1955. *Leptoguignardia*, eine neue Gattung der bitunicaten Ascomyceten. Sydowia **9**: 216-220.

——. 1957. Haupt- und Nebenfruchtformen bei *Guignardia* Viala et Ravez. Sydowia **11**: 210-224.

——. 1959. Über drei *Guignardia*-Arten und ihre Nebenfruchtformen. Phytopath. Z. **34**: 411-416.

——. 1963. *Epibelonium*, eine neue Gattung der Schizotheriaceen (Ascomyceten). Phytopath. Z. **47**: 239-243.

—— **& J. J. A. von Arx.** 1950. Einige Aspekte zur Systematik pseudosphaerialer Ascomyceten. Ber. Schweiz. Bot. Ges. **60**: 329-395.

—— & ——. 1962. Die Gattungen der didymosporen Pyrenomyceten. Beitr. Kryptfl. Schweiz. **11**(2): 1-922.

—— **& E. G. Baumeister.** 1957. Kulturversuche mit drei *Cucurbitaria*-Arten. Sydowia **11**: 70-75.

—— **& W. Loeffler.** 1957. Über die Gattung *Clathridium* (Sacc.) Berl. Sydowia **11**: 116-120.

—— **& R. Menon.** 1954. *Bagcheea*, eine neue Gattung der Sphaeriales aus Indien. Phytopath. Z. **22**: 417-420.

Munk, A. 1953. The system of the Pyrenomycetes. Dansk Bot. Arkiv **15**(2): 1-163.

——. 1957. Danish Pyrenomycetes. A preliminary Flora. Dansk Bot. Arkiv **17**: 1-491.

Nannfeldt, J. A. 1932. Studien über die Morphologie und Systematik der nichtlichenisierten inoperculaten Discomyceten. Nova Acta Reg. Soc. Sci. Uppsaliensis ser. IV, **8**(2): 1-368.

Nelson, R. R. 1957. Heterothallism in *Helminthosporium maydis*. Phytopathology **47**. 191-192.

——. 1959. A major gene locus for compatability in *Trichometasphaeria turcica.* Phytopathology **49**: 159-160.

——. 1959a. *Cochliobolus carbonum* the perfect stage of *Helminthosporium carbonum.* Phytopathology **49**: 807-811.

——. 1960. Evolution of sexuality and pathogenicity. I. Interspecific crosses in the genus *Helminthosporium.* Phytopathology **50**: 375-377.

——. 1961. Evolution of sexuality and pathogenicity. II. A comparison of the pattern of sexuality in *Cochliobolus victoriae* and related species. Phytopathology **51**: 222-223.

——. 1965. The genetic control of mating types in *Cochliobolus lunatus* and *Cochliobolus intermedia.* Mycologia **57**: 984-986.

Nienburg, W. 1914. Zur Entwicklungsgeschichte von *Polystigma rubrum* DC. Zeit. Bot. **6**: 369-400.

Niessl, G. von. 1875. Neue Kernpilze. I Serie. Österr. Bot. Z. **25**: 46-49, 85-87, 165-166 199-203.

Noack, F. 1905. *Helminthosporium gramineum* Rabenh. und *Pleospora trichostoma* Wint. Z. Pflznkr. **15**: 193-205.

Obrist, W. 1959. Untersuchungen über einige "Dothidiale" Gattungen. Phytopath. Z. **35**: 357-388.

Olive, E. W. 1905. The morphology of *Monascus purpureus.* Bot. Gaz. **39**: 56.

Orton, C. R. 1924. Studies in the morphology of the Ascomycetes. I. The stroma and the compound fructifications of the Dothideaceae and other groups. Mycologia **16**: 49-95.

——. 1944. Graminicolous species of *Phyllachora* in North America. Mycologia **36**: 18-53.

Parker, A. K. 1957. *Europhium*, a new genus of the Ascomycetes with a *Leptographium* imperfect state. Canad. J. Bot. **35**: 173-179.

Pelletier, R. L. & G. W. Keitt. 1954. *Venturia inaequalis* (Cke.) Wint. VI. Amino-acids as sources of nitrogen. Amer. J. Bot. **41**: 362-371.

Persoon, C. H. 1801. Synopsis Methodica Fungorum. 706 pp. Goettingae, Dieterich.

Petrak, F. von. 1924. Mykologische Notizen. VII. Ann. Mycol. **22**: 1-182.

——. 1928. Über *Bagnisiopsis* und verwandte Gattungen. Hedwigia **68**: 251-290.

——. 1928a. Über *Englerula* und die Englerulaceen. Ann. Mycol. **26**: 385-413.

——. 1941. Mykologische Notizen. XIV. Ann. Mycol. **39**: 251-349.

——. 1966. Über die Gattung *Cryptospora.* Sydowia **19**: 268-279.

Pomerleau, R. 1938. Recherches sur le *Gnomonia ulmea* (Schw.) Thum. Naturaliste Canadien Contr. Bot. Univ. Montreal **31**: 1-139.

Pontecorvo, G. & J. A. Roper. 1952. Genetic analysis without sexual reproduction by means of polyploidy in *Aspergillus nidulans.* J. Gen. Microbiol. **6**: vii. (Abst.)

—— & ——. 1956. The parasexual cycle in fungi. Ann. Rev. Microbiol. **10**: 303-400.

Povah, A. 1924. *Hypoxylon* poplar canker. Phytopathology **14**: 140-145.

Raper, K. B. & C. Thom. 1949. A Manual of the Penicillia. The Williams & Wilkins Co., Baltimore.

Ray, W. W. 1936. Pathogenicity and cultural experiments with *Caliciopsis pinea.* Mycologia **28**: 201-208.

Rehm, H. 1896. Discomycetes. *In* Rabenhorst's "Kryptogamenflora von Deutschland, Oesterreich und der Schweiz." Abt. III, **1**(3): 1-1275.

Reid, J. & C. Booth. 1966. The genus *Flageoletia* (Sacc.) Höhn. Canad. J. Bot. **44**: 445-451.

Reusser, F. A. 1964. Über einige Arten der Gattung *Guignardia* Viala et Ravaz. Phytopath. Z. **51**: 205-240.

Riedl, H. 1959. Kulturversuche zum Pleomorphismus einiger Pyrenomyceten. Oesterr. Bot. Z. **106**: 478-545.

Robbins, W. J. 1950. A survey of the growth requirements of some Basidiomycetes. Mycologia **42**: 470-476.

Rogerson, C. T. 1970. The Hypocrealean Fungi (Ascomycetes, Hypocreales). Mycologia **62**: 865-910.

Ryan, R. W. 1926. The development of the perithecia in the Microthyriaceae and comparison with *Meliola.* Mycologia **18**: 100-110.

Saccardo, P. A. 1882-1931. Sylloge Fungorum. 1-25.

Sachs, J. 1874. Lehrbuch der Botanik. Vierte Auflage. Wilhelm Engelmann, Leipzig.

——. 1896. Physiologische Notizen. X. Phylogenetische Aphorismen und über innere Gestaltungsursachen oder Automorphen. Flora **82**: 173-223.

Salmon, E. S. 1900. A monograph of the Erysiphaceae. Mem. Torrey Bot. Club. **9**: 1-292.

Sartoris, G. B. & C. H. Kauffman. 1925. The development and taxonomic position of *Apiosporella collinsii.* Pap. Mich. Acad. Sci. **5**: 149-162.

Scheinpflug, H. 1958. Untersuchungen über die Gattung *Didymosphaeria* Fuck. und einige verwandte Gattungen. Ber. Schweiz. Bot. Ges. **68**: 325-385.

Schopfer, W. H. 1954. Les facteurs de croissance pour les microorganismes. VIII. Congr. Int. Bot. Rapp. et Commun. Sect. **18-20**: 103-110.

Seaver, F. J. 1910. Hypocreales. N. Amer. Flora **3**: 1-56.

——. 1909-11. The Hypocreales of North America. Mycologia **1**: 41-75; 177-207; **2**: 48-92; **3**: 207-230.

Seeler, E. V. 1940. The diseases of *Gleditsia* caused by a species of *Thyronectria*. J. Arnold Arb. **21**: 405-427.

——. 1940a. A monographic study of the genus *Thyronectria*. J. Arnold Arb. **21**: 429-460.

——. 1943. Several fungicolous fungi. Farlowia **1**: 119-133.

Shanor, L. 1952. The characteristics and morphology of a new genus of the Laboulbeniales on an earwig. Amer. J. Bot. **39**: 498-504.

Shear, C. L. & B. O. Dodge. 1922. Life history of an undescribed ascomycete isolated from a granular mycetoma of man. Mycologia **14**: 239-243.

—— & ——. 1927. Life histories and heterothallism of the red bread mold fungi of the *Monilia sitophila* group. J. Agr. Res. **34**: 1019-1042.

—— **& N. E. Stevens.** 1917. *Endothia parasitica* and related species. U. S. Dept. Agr. Bull. **380**: 1-82.

—— & ——. 1924. *Botryosphaeria* and *Physolospora* on currant and apple. J. Agr. Res. **28**: 589-598.

Shirakawa, H. S. 1956. The development of *Coccostroma* (*Bagnisiopsis*) *nuda*. Amer. J. Bot. **43**: 372-377.

Shoemaker, R. 1955. Biology, cytology and taxonomy of *Cochliobolus sativus*. Canad. J. Bot. **23**: 362-376.

——. 1964. Revision of some *Dimeriella* and *Dimerosporium* parasites on conifers. Canad. J. Bot. **43**: 631-639.

—— **& E Müller**. 1963. Generic correlations and concepts: *Broomella* and *Pestalotia*. Canad. J. Bot. **41**: 1235-1243.

Snyder, W. C. 1940. White perithecia and the taxonomy of *Hypomyces ipomoeae*. Mycologia **32**: 646-648.

—— **& H. N. Hansen.** 1940. The species concept of *Fusarium*. Amer. J. Bot. **27**: 64-67.

—— & ——. 1941. The species concept in *Fusarium* with reference to section *Martiella*. Amer. J. Bot. **28**: 738-742.

—— & ——. 1945. The species concept in *Fusarium* with reference to *Discolor* and other sections. Amer. J. Bot. **32**: 657-666.

Solms-Laubach, H. G. 1887. *Penicilliopsis clavariaeformis*, eine neuer javanischer Ascomycet. Ann. Jard. Buitenzorg **6**: 53-72.

Spaulding, P., T. J. Grant & T. T. Ayers. 1936. Investigations of *Nectria* diseases in hardwoods of New England. J. For. **34**: 169-179.

Sprague, R. & F. D. Heald. 1927. A witches broom of service-berry. Trans. Amer. Micro. Soc. **46**: 219-247.

Srb, A. M. 1950. Complex growth requirement of an arginine-less mutant of *Neurospora*. Bot. Gaz. **111**: 470-476.

—— **& N. H. Horowitz.** 1944. The ornithine cycle in *Neurospora* and its genetic control. J. Biol. Chem. **154**: 129-139.

Starback, K. 1899. Ascomyceten der ersten Regnellschen Expedition I. Bihang K. Svensk. Vetenska. Akad. Handl. 25.

Stelling-Dekker, N. M. 1931. Die sporogenen Hefen. Verhandl. Akad. Wettensch-Amsterdam, Afdeel. Natuurk, 2nd ser. **28**: 1-575.

Stevens, F. L. 1918. Some meliolicolous parasites and commensals from Porto Rico. Bot. Gaz. **65**: 227-248.

——. 1927. The Meliolineae. I. Ann. Mycol. **25**: 405-469.

——. 1928. The Meliolineae. II. Ann. Mycol. **26**: 165-383.

——. 1934. *Physalospora malorum* on currant. J. Agr. Res. **28**: 583-587.

——. 1936. Two species of *Physalospora* in England. Mycologia **28**: 330-336.

Stolk, A. C. 1955. *Emericellopsis minima* sp. nov. and *Westerdykella ornata* gen nov., sp. nov. Trans. Brit. Mycol. Soc. **38**: 419-434.

——. 1955a. The genera *Anixiopsis* Hansen and *Pseudeurotium* van Beyma. Anton Leeuwenhoek **21**: 65-78.

Stoppel, R. 1907. *Eremascus fertilis* nov. sp. Flora **97**: 332-346.

Thaxter, R. 1896. Contributions towards a monograph of the Laboulbeniaceae. I. Mem. Amer. Acad. Arts & Sci. **12**: 189-429.

——. 1908. Contributions towards a monograph of the Laboulbeniaceae. II. Mem. Amer. Acad. Arts & Sci. **13**: 217-469.

——. 1924. Contributions towards a monograph of the Laboulbeniaceae. III. Mem. Amer. Acad. Arts & Sci. **14**: 309-426.

——. 1926. Contributions towards a monograph of the Laboulbeniaceae. IV. Mem. Amer. Acad. Arts & Sci. **15**: 427-589.

——. 1931. Contributions towards a monograph of the Laboubeniaceae. V. Mem. Amer. Acad. Arts & Sci. **16**: 1-435.

Theissen, F. 1913. Hemisphaeriales. Ann. Mycol. **11**: 468-469.

——. 1916. Beiträge zur Systematik der Ascomyzeten. Ann. Mycol. **14**: 401-439.

—— **& H. Sydow.** 1915. Die Dothideales. Ann. Mycol. **13**: 149-430.

—— **&** ——. 1917. Synoptische Tafeln. Ann. Mycol. **15**: 389-396.

—— **&** ——. 1918. Vorentwürfe zu den Pseudosphaeriales. Ann. Mycol. **16**: 1-34.

Thom, C. & M. B. Church. 1926. The Aspergilli. 212 pp. Williams & Wilkins, Baltimore.

Tilak, S. T. 1963. The genus *Parodiella* in Bombay. J. Indian Bot. Soc. **42**: 110-117.

Tinline, R. D. 1951. Studies on the perfect stage of *Helminthosporium sativum*. Cand. J. Bot. **29**: 467-478.

——. 1954. Physiology, pathogenicity and genetics of *Helminthosporium sativum*. Thesis. Univ. Wisconsin.

—— **& J. G. Dickson.** 1958. *Cochliobolus sativus*. I. Perithecial development and the inheritance of spore color and mating types. Mycologia **50**: 697-706.

Togaschi, K. 1924. Some studies on a Japanese apple canker and its causal fungus, *Valsa mali*. J. Col. Agric. Hokkaido Imp. Univ. **12**: 265-324.

——. 1930. Comparative studies on the physiology of *Leucostoma leucostoma* and *Valsa japonica*. Bull. Imp. Coll. Atr. For. Morioka, Japan No. 15. 76 pp.

——. 1930a. Morphological studies of *Leucostoma leucostoma* and *Valsa japonica*, the causal fungi of canker or die-back disease of peach trees. Bull. Imp. Coll. Agr. For. Morioka, Japan No. 14. 50 pp.

——. 1931. Studies on the pathology of peach canker. Bull. Imp. Coll. Agr. For. Morioka, Japan. No. 16. 178 pp.

Traverso, G. B. 1907. Flora italica Cryptogama, Pyrenomycetineae. Pt. I, fasc. **2**: 420-470. Rocca S. Càsciano.

Urban, Z. 1957. Ceskoslovenskych zastupcu rodu *Valsa leucostoma* a *Valsalla* Rezpravy. Ceszkoslovenske Akad. Ved. **68**: 1-100.

Urries, M. J. de. 1946. Estudio citologico y experimental de *Leptosphaeria cavanillesii* nov sp. Ann. Jard. Bot. Madrid.

Van Beyma, F. H. 1937. Beschreibung einiger neuer Pilzarten aus dem Centraalbureau voor Schimmelcultures, Baarn (Holland). IV. Mitteil. Zentralbl. Bakt. Parasitenk. Infectionskr. **96**: 411-422.

Van Buren, G. 1915. Die schweizerischen Protomyceten mit besonderer Berücksichtigung ihrer Entwicklungsgeschichte und Biologie. Beitr. Krypt.-Fl. Schweiz **5**(1): 1-95.

——. 1922. Weitere Untersuchungen über die Entwicklungsgeschichte und Biologie der Protomycetaceen. Beitr. Krypto.-F. Schweiz **5**(3): 1-94.

Varitchak, B. 1928. L'evolution nucleare chez *Ascoidea rubescens* Brefeld. Compt. Rend. Acad. Sci. Paris **186**: 96-98.

Walker, L. B. 1931. Studies on *Ascoidea rubescens.* I. History and development. Mycologia **23**: 51-76.

Ward, H. M. 1883. On the morphology and development of the perithecium of *Meliola*. Proc. Roy. Soc. London **34**: 388-390.

——. 1883a. On the morphology and development of the perithecia of *Meliola*, a genus of tropical epiphyllous fungi. Phil. Trans. Roy. Soc. London **174**: 583-598.

Waterman, A. M. 1945. The blight of species of *Abies* caused by species of *Rehmiellopsis*. J. Agr. Res. **70**: 315-337.

Webster, J. 1951. Graminicolous pyrenomycetes. II. The occurrence of the perfect stage of *Helminthosporium teres* in Britain. Trans. Brit. Mycol. Soc. **34**: 309-317.

——. 1955. Graminicolous pyrenomycetes. V. Conidial states of *Leptosphaeria michotii, L. microscopica, Pleospora vagans* and the perfect state of *Dinemasporium graminum*. Trans. Brit. Mycol. Soc. **38**: 347-365.

——. 1955a. *Hendersonia typhae* the conidial state of *Leptosphaeria typharum*. Trans. Brit. Mycol. Soc. **38**: 405-408.

——. 1956. Conidia of *Acrospermum compressum* and *A. graminum*. Trans. Brit. Mycol. Soc. **39**: 361-366,

——. 1957. *Pleospora straminis, P. rubelloides* and *P. rubicunda*; three fungi causing purple staining of decaying tissue. Trans. Brit. Mycol. Soc. **40**: 177-186.

—— **& H. H. Hudson.** 1957. Graminicolous pyrenomycetes VI. Conidia of *Ophiobolus herpotrichus, Leptosphaeria lactuosa, L. fuckelii, L. pontiformis* and *L. eustomoides*. Trans. Brit. Mycol. Soc. **40**: 509-522.

—— **& M. T. Lucas.** 1959. Observations on British species of *Pleospora*. I. Trans. Brit. Mycol. Soc. **42**: 332-342.

—— **&** ——. 1961. Observations on British species of *Pleospora* II. Trans. Brit. Mycol. Soc. **44**: 417-436.

Wehmeyer, L. E. 1925. Cultural life histories of certain species of *Eutypella, Diatrypella* and *Cryptovalsa*. Pap. Mich. Acad. Sci. **5**: 179-194.

——. 1926. A biologic and phylogenetic study of the stromatic Sphaeriales. Amer. J. Bot. **13**: 575-645.

——. 1933. The genus *Diaporthe* and its segregates. Univ. Mich. Press. Sci. ser. **9**: 1-349.

——. 1941. *Pseudotrichia* and the new genus *Phragmodiaporthe*. Mycologia **33**: 54-63.

——. 1941a. A revision of *Melanconis, Pseudovalsa, Prosthecium* and *Titania*. Univ. Mich. Press. Sci. ser. **14**: 1-161.

——. 1941b. The genus *Thyridaria* (Pyrenomycetes). Lloydia **4**: 241-261.

——. 1954. Perithecial development in *Pleospora trichostoma*. Bot. Gaz. **115**: 297-310.

——. 1955. Development of ascostroma in *Pleospora armeriae* of the *Pleospora herbarum* complex. Mycologia **47**: 821-834.

——. 1961. A world monograph of the genus *Pleospora* and its segregates. Univ. Mich. Press. Ann Arbor. 451 pp.

Welch, D. S. 1926. A monographic study of the genus *Cucurbitaria* in North America. Mycologia **18**: 50-87.

Wells, D. E. 1954. *Ascocybe*, a new genus of lower ascomycetes. Mycologia **46**: 37-51.

Wheeler, H. E. 1950. Genetics of *Glomerella* VIII. The genetic basis of the occurrence of minus mutants. Amer. J. Bot. **37**: 304-312.

—— **& C. H. Driver.** 1953. Genetics and cytology of a mutant dwarf-spored *Glomerella*. Amer. J. Bot. **40**: 694-702.

—— **& J. W. McGahen.** 1952. Genetics of *Glomerella*. X. Genes affecting sexual reproduction. Amer. J. Bot. **39**: 110-119.

——**, L. S. Olive, C. T. Ernest & C. W. Edgerton.** 1948. Genetics of *Glomerella*. V. Crozier and ascus development. Amer. J. Bot. **35**: 722-728.

White, W. L. 1942. A new Hemiascomycete. Canad. J. Res. C, **20**: 389-395.

Wingard, S. A. 1922. Yeast-spot of lima beans. Phytopathology **12**: 525-532.

——. 1925. Studies on the pathogenicity, morphology and cytology of *Nematospora phaseoli*. Bull. Torrey Bot. Club. **52**: 249-290.

Winge, O. 1911. Encore le *Sphaerotheca castagnei*. Bull. Soc. Mycol. Fr. **27**: 211.

——. 1935. On haplophase and diplophase in some Saccharomycetes. Compt. Rend. Lab. Carlsberg, ser. physiol. **21**: 77-108.

——. 1944. On segregation and mutation in yeasts. Compt. Rend. Lab. Carlsberg. ser. physiol. **24**: 79-95.

——. 1947. The segregation in the ascus of *Saccharomycodes ludwigii*. Compt. Rend. Lab. Carlsberg, ser. physiol. **24**: 223-226.

——. 1951. The relation between yeast cytology and genetics. Compt. Rend. Lab. Carlsberg, ser. physiol. **25**: 85-97.

——. 1952. The genetic situation concerning fermentation in yeasts. Heredity **6**: 263-269.

—— **& O. Lausten.** 1937. On two types of spore germination and on genetic segregations in *Saccharomyces*, demonstrated through single spore culture. Compt. Rend. Lab. Carlsberg, ser. physiol. **22**: 99-116.

—— & ——. 1938. Artificial species hybridization in yeast. Compt. Rend. Lab. Crlsberg. ser. physiol. **22**: 245-247.

—— & ——. 1939. On 14 new yeast types, produced by hybridization. Compt. Rend. Lab. Carlsberg, ser. physiol. **22**: 337-352.

—— & ——. 1939a. *Saccharomycodes ludwigii* Hansen, a balanced heterozygote. Compt. Rend. Lab. Carlsberg, ser. physiol. **22**: 357-370.

—— **& C. Roberts.** 1950. The polymeric genes for maltose fermentation in yeasts and their mutability. Compt. Rend. Lab. Carlsberg, ser. physiol. **25**: 36-83.

—— & ——. 1952. The relation between genes for maltose, raffinose and suberose fermentation in yeasts. Compt. Rend. Lab. Carlsberg, ser. physiol. **25**: 141-171.

—— & ——. 1953. The genes for maltose and raffinose fermentation in *Saccharomyces cerevisiae*. Compt. Rend. Lab. Carlsberg, ser. physiol. **25**: 241-257.

—— & ——. 1954. Causes of deviation from the 2:2 segregation in tetrads of monohybrid yeasts. Compt. Rend. Lab. Carlsberg, ser. physiol. **25**: 285-329.

—— & ——. 1955. Identification of the maltose genes in some American haploid and European diploid yeasts. Compt. Rend. Lab. Carlsberg, ser. physiol. **25**: 331-340.

—— & ——. 1957. A genetic analysis of melibiose and raffinose fermentation. Compt. Rend. Lab. Carlsberg, ser. physiol. **25**: 419-450.

Winter, G. 1887. Ascomyceten. *In* Rabenhorst's Die Pilze Deutschlands, Oesterreichs und der Schweiz. Bd. I, Abt. **II**: 1-928.

Wolf, F. A. 1935. Morphology of *Polythrincium*, causing sooty blotch of clover. Mycologia **27**: 58-73.

——. 1935a. The perfect stage of *Cercospora rubi*. Mycologia **27**: 347-356.

——. 1936. False mildew of red mulberry. Mycologia **28**: 268-277.

——. 1938. Life histories of two leaf inhabiting fungi on sycamore. Mycologia **30**: 54-63.

——. 1938a. *Cercospora* leaf spot of rose caused by *Mycosphaerella rosicola*. Mycologia **30**: 282-298.

——. 1940. A leaf spot ot *Nyssa*. Mycologia **32**: 331-335.

——. 1943. The perfect stage of *Cercospora sordida*. Mycologia **35**: 503-509.

—— **& W. J. Barbour.** 1941. Brown spot needle disease of pines. Phytopathology **31**: 61-74.

Wollenweber, H. W. 1913. *Ramularia, Mycosphaerella, Nectria, Calonectria.* Eine morphologisch-pathologische Studie zur Abgrenzung von Pilzgruppen mit cylindrischen und sichelförmigen Conidien. Phytopathology **3**: 197-241.

——. 1926. Pyrenomyceten Studien. I. Angew. Bot. **6**: 300-313; **8**: 168-212.

—— **& O. A. Reinking.** 1935. Die Fusarien. 355 pp. Berlin.

Yarwood, C. E. 1935. Heterothallism of sunflower powdery mildew. Science. N. S. **82**: 417-418.

Zalazky, H. 1964. Microconidial state of *Cucurbitaria staphula* Dearness ex R. H. Arnold & R. S. Russell. Canad. J. Bot. **42**: 1586-1588.

Zender, J. 1925. Sur quelques nouvelles espèces de levures et d'*Endomyces*. Bull. Soc. Bot. Geneve II, **17**: 258-271.

——. 1925a. Sur les classification des Endomycetaceae. Bull. Soc. Bot. Geneve II, **17**: 272-302.

Zogg, H. 1949. Beiträge zur Kenntnis der brasilianischen Hysteriaceen. Ber. Schweiz. Bot. Ges. **59**: 23-44.

——. 1960. Über die Gattungen *Actidium* Fries. und *Bulliardiella* (Sacc.) Paoli. Ber. Schweiz. Bot. Ges. **70**: 195-205.

——. 1962. Die Hysteriaceae s. str. und Lophiaceae. Beitr. Krypt.-Fl. Schweiz. **11**: 1-190.

INDEX